AF702192

Thomas Jefferson, Time, and History

Jeffersonian America

Jan Ellen Lewis, Peter S. Onuf, and Andrew O'Shaughnessy, Editors

THOMAS JEFFERSON, TIME, *and* HISTORY

Hannah Spahn

University of Virginia Press *Charlottesville and London*

University of Virginia Press
© 2011 by the Rector and Visitors of the University of Virginia
All rights reserved
Printed in the United States of America on acid-free paper

First published 2011

1 3 5 7 9 8 6 4 2

LIBRARY OF CONGRESS CATALOGING-IN-PUBLICATION DATA
Spahn, Hannah, 1976-
Thomas Jefferson, time, and history / Hannah Spahn.
p. cm. – (Jeffersonian America)
Includes bibliographical references and index.
ISBN 978-0-8139-3168-5 (cloth : alk. paper) – ISBN 978-0-8139-3204-0 (e-book)
1. Jefferson, Thomas, 1743-1826–Philosophy. 2. Time–Philosophy.
3. History–Philosophy. I. Title.
B885.Z7S63 2011
115.092–dc22

2011011980

MONTICELLO

Preparation of this volume has been supported by
the Thomas Jefferson Foundation.

Für Nicolo und Niklas

Contents

Acknowledgments

This book would not have been possible without a number of individuals and institutions in Europe and the United States. I am very grateful, first of all, to my teachers at the John F. Kennedy Institute for North American Studies of the Free University Berlin, my *Doktorvater,* Winfried Fluck, and Heinz Ickstadt, for their advice and guidance. Many thanks also to Ulla Haselstein and Manfred Pfister for their support. In general, I have benefited greatly from the stimulating intellectual environment at the Kennedy Institute and from the helpful suggestions I received at various stages of the work from friends and colleagues, including Stefan Brandt, Steffi Dippold, Andrew Gross, Edda Luckas, Frank Mehring, Roxana Preda, Simon Schleusener, Mary Ann Snyder-Körber, and Kirsten Twelbeck.

Over the years, I have also incurred intellectual debts at other European universities. I would like to thank various scholars for productive conversations and criticism, especially Barbara Buchenau, Ari Helo, Monica Henry, Frank Kelleter, Susan Manning, Peter Nicolaisen, Marie-Jeanne Rossignol, Dietmar Schloss, Rainer Schnoor, Eran Shalev, Barbara Stollberg-Rilinger, Maurizio Valsania, and Naomi Wulf.

The dissertation was funded by a FAZIT Foundation fellowship, a fellowship from the Federal State of Berlin, a travel fellowship to take part in the Young Scholars Forum at the German Historical Institute in Washington, DC, and a two-month research fellowship from the Robert H. Smith International Center for Jefferson Studies in Charlottesville, Virginia. I was very fortunate later on to be able to return to the ICJS as a Gilder Lehrman

Fellow for nine months to do research and work on the manuscript for this book.

In the beautiful environment of Jefferson's "magic mountain" I found ideal working conditions, owing to the extremely friendly and competent staff of the Thomas Jefferson Foundation and to the community of early Americanists meeting in the colloquia of the ICJS. For their constructive criticism I thank especially Christa Dierksheide, Adam Jortner, J. Jefferson Looney, Johann Neem, Kate Pierce, Sophia Rosenfeld, Leonard Sadosky, Paula Viterbo, Gaye Wilson, and Nicholas Wood. In the final stages of work on the book, moreover, I benefited from the precision of Joanne Allen's copyediting and the assistance especially of Dick Holway and Raennah Mitchell at the University of Virginia Press.

Among the Jeffersonians, I am particularly grateful to three individuals, Andrew O'Shaughnessy, Frank Cogliano, and the *paterfamilias* of Jefferson's intellectual *oikos* today, Peter Onuf. I thank them for their generous support over many years, for making me feel welcome in the world of Jefferson research, and for taking such a close interest in my work and improving it greatly by their unique scholarship and fellowship.

Finally, many thanks to Joanne and Richard Fishbane for introducing me to the fascination of American literature and history during my first stay in the United States twenty years ago. I am deeply grateful to my parents, Renate and Peter Spahn, for introducing me to the foreign country of the past more generally and to my sister, Barbara Spahn, for being such a reliable (and fun) traveling companion there. To my husband, Nicolo Unger, I am indebted for many things, not the least being his generous willingness to look interested despite (for a while) being treated to at least one detailed monologue about Thomas Jefferson a day. My special thanks to our baby son, Niklas Jonathan, in particular for sleeping (sometimes) and smiling (often) during the final months of work on the book.

Thomas Jefferson, Time, and History

Introduction

"The documents of your childhood, your letters, correspondencies, notes, books, &c., &c., all gone! And your life cut in two, as it were, and a new one to begin, without any records of the former." Thus Thomas Jefferson sympathized with his favorite granddaughter, Ellen Randolph Coolidge, when news came that the ship carrying her baggage had sunk en route to Boston, the home of her new husband. Her belongings had included not only written documents but also a writing desk crafted by the Monticello slave John Hemings. The craftsman, Jefferson wrote, now joined her in grief over its loss: "Virgil could not have been more afflicted had his Aeneid fallen a prey to the flames." As a consolation and substitute, Jefferson sent Ellen and her husband a "writing box" that he had designed himself half a century earlier, "not claiming the same value from it's decorations, but from the part it has *borne* in our history and the events with which it has been associated." Comparing its future public function to that of "relics of the church," he was convinced that its "imaginary value would increase with the years." The unique gift that Jefferson announced to his granddaughter in Massachusetts was the portable desk on which he had drafted the Declaration of Independence.[1]

This gesture in the last year of Jefferson's life illustrates important aspects of his conception of time and his vision of American history. First, his replacement of Ellen's writing desk with the desk on which he had drafted the founding document of the republic reveals an attitude toward the past that fused the personal and the political. The gesture also blended this individual and national history–his granddaughter's lost correspondence and the draft of the Declaration of Independence–with the means and the relics of its construction: in his reference to John Hemings's "Aeneid" (a reference whose patronizing overstatement, from today's perspective, all but conceals how much of America's founding history actually rested on the work of American slaves) Jefferson went as far as to identify the device of the writing desk with a national epic about the founding of an empire. Finally, Jefferson's letter testifies to a peculiar temporality, one in which the present did not flow smoothly from the past but was sharply dissociated from it by violent events that "cut" life "in two"–in Ellen's personal history, by the shipwreck accompanying her wedding journey, and in national American history, as Jefferson saw it, by the "last stab to agonizing affection" irretrievably separating the United States from Europe in his draft of the Declaration,[2] or by the impregnable line between the living and the dead that defined his conception of generational sovereignty.[3] Jefferson's efforts to construct a narrative of American progress were premised on such moments of radical discontinuity with the past. More than many of his contemporaries, he tended to insist on their necessity, while also recognizing and even lamenting their alienating effects. As I shall argue, this ambivalence toward the modern experience of breaking with the old pervades Jefferson's writings. Thus, his influential celebration of American "newness"[4] did not keep him from making claims to the extreme "antiquity"[5] of American society. And there is hardly a text by this famously "forward-looking"[6] visionary of future progress that is not, in fact, deeply concerned with looking back to individual or collective history. As Jefferson's case illustrates, the break with the past in the Enlightenment was closely intertwined with the emergence of a new appreciation of the past for its own sake.

This book is about Thomas Jefferson's Janus-faced attitude toward what the modernist chronicler of Jefferson, Mississippi, called "the fluid cradle of events (time)."[7] Unlike the creator of the fictional Jefferson, the historical Thomas Jefferson approached time mainly by trying to diminish its "fluid" character. His imaginative efforts to tame the flow of time were, however, far from uncomplicated. In Jefferson's endeavor to structure time by dividing it into clear-cut stages, he was continually confronted with, and often inclined to accept, essential intellectual and emotional impediments. The roots of this divided experience were quite specific. It is a contention of the following chapters that Jefferson's attempts to come to terms with the decisive problem of historical change in a period of revolutions cannot be explained without an analysis of his Newtonian conception of time. "Newtonianism," as understood here, was a transitional worldview on the brink of modernity. Its enlightened proponents provided the groundwork for modern philosophy mostly without being able to cross the line themselves. They lived in a universe in which "*time,* which reveals all other things," as John Locke put it,[8] was just beginning to assume its modern significance. In the eighteenth century, eternity was only in the process of being secularized, or to put it another way, "temporalized": Isaac Newton had already formulated the modern idea of a universal and all-encompassing "absolute" time, but the full implications of this foundational secularization had yet to be realized, in the natural sciences as well as in explanations of historical change.[9] The result was a Newtonian culture of approximation and compromise, a culture that remained skeptical of secular absolutes and tended to display a modest awareness of its own circumstantiality.[10] Even the most "progressive" eighteenth-century minds thus took for granted, far more than the Romantic image of the Enlightenment's philosophical hubris seems to suggest in retrospect, fundamental human limitations in the pursuit of secular progress.[11]

Jefferson, as I see him, was a consummate product of this Newtonian culture. The self-consciously approximative character of his empiricism extended even to the details of his approach to time. He typically allowed himself to draw by eye, whether in the freehand constructions of his clocks at Monticello[12] or in the idiosyncratic temporal contours of his argument for generational sovereignty. In this complex reflection on the political significance of time, for instance, he combined a recourse to specific temporal data from natural history with a rather lax application of them. Thus, he began his generational calculations by carefully drawing the precise figure of eighteen years and eight months from the mortality tables established by the Count de Buffon. However, he did not fully rely on the work of this authoritative natural historian for the timing of each generation. Nor did he go as far as to use the more convenient time span of two full decades in his political conclusions. Instead, Jefferson did something that was characteristic of his mode of thinking in general: he chose the middle way between natural and civil history by rounding Buffon's eighteen years and eight months off to the still somewhat crooked figure of nineteen years. What may look today like a somewhat capricious method, or indeed a lack of method, was less an accident or a personal quirk than the consistent expression of a Newtonian Enlightenment in which all human endeavors were considered to be essentially approximative, always a compromise between the naturally given and the historically attained.

The self-consciously provisional nature of all human knowledge and agency in this worldview had important moral and political implications. Ultimately, it was responsible for Jefferson's inclination to reconcile in his arguments what often strike us, in retrospect, as contradictory positions. In Jefferson's own time, these "Jeffersonian" antinomies (sometimes identified as typically "American" antinomies as well) were not perceived as such to the same extent. They were part of a mentality that was philosophically bent on compromise, a mentality that made it comparatively easy for men like Jefferson to argue for liberty in a language of compulsion, to link the pain of separation to a celebration of independence, or to think of themselves as cosmopolitan nationalists, farming admirers of the steam engine, and slaveholding opponents of slavery. In a culture that remained, at bottom, noncommital in its answers to a crucial question defining the human condition–the nature of change in time–the American founding generation could begin to shape the modern world without conceiving of it consistently in modern terms.

Jefferson's writings derive their philosophical relevance and their contradictory beauty from the fact that while he sought to emulate his cultural ideal of a versatile, open-minded Enlightenment, he was becoming sensitive to the intellectual tensions it created. More, perhaps, than many of his peers, he was intent on relieving these tensions; at the same time, he sought to remain within the paradoxical confines of his ideal. This dual orientation of Jefferson's thought is keenly evident in his ambivalent efforts to control the "fluidity" of change in a conspicuously changing world. Its most important result may have been his contribution to a narrative that is often summed up as America's exceptional "escape" from time and history, an exceptionalist interpretation much admired, for instance, in my own country since the days of Jefferson's contemporary Goethe.[13] The evolution of this exceptionalism in Jefferson's writings had far-reaching moral and political consequences. As I shall argue, it was closely tied both to his changing views on the French Revolution and to a fateful shift in his approach to slavery.[14] Whether Jefferson and his contemporaries expected Americans to "escape" from time into space and back,[15] or rather from one temporality into another (as I claim in this book), their imagined flight was always a highly ambivalent construction. Intriguingly, as Jefferson's writings illustrate, it was accompanied not only by expressions of disdain for history but also by the contrary phenomenon of a passionate appreciation of it, especially in the rhetorical creation of a new, uniquely American past.

Jefferson's generation was instrumental in founding, along with the political institutions of the United States, a lasting culture of national memory. In his efforts to construct a progressive narrative about America's national origins, Jefferson typically sought to have it both ways, trying to couple dramatic ruptures with the creation of historical relics that furnished new reasons for looking back. Without ever crossing the "divide" into a fully historicist conception of history,[16] his writings were oriented both forward and backward, extending his enlightened view of the past far into the future. Thus, Jefferson's construction of new historical continuities has had an important impact on American culture, most conspicuously concerning the image of the founding generation itself. In their transitional Newtonian worldview, Jefferson and other founders found a mode of self-historicization that has persisted in the following centuries. As a consequence, it has been particularly difficult for scholars to achieve the double goal of understanding the American revolutionaries within their self-created historical comfort zone while also avoiding the danger of reiterating, in one way or another, their self-serving historical arguments. By analyzing the conceptions of time and history that constituted Jefferson's approach to revolutionary change, this book seeks to contribute to this complicated but necessary endeavor.

The Founders in the Praesens Historicum

It appears that one of the lasting achievements of the American founding generation was that they managed to establish special relations to time and history. Owing to a complex array of factors, the founders were extraordinarily effective in forging close connections to their posterity. Their posthumous existence today is far more intense, for instance, than that of their contemporaries in European history: few people would care to know what other "great" eighteenth-century figures, such as Frederick II of Prussia or Catherine II of Russia, would do or say if they were miraculously brought back to life today. By contrast, the opinions of the American revolutionaries continue to be solicited in contemporary discussions, and their politics are directly compared with recent political trends, almost as if they were still living elder statesmen. Like the classical *praesens historicum,* appreciated by the founders in their ancient reading, this slightly surreal temporal phenomenon of the founders' enduring presence in American culture often has the effect of provoking not only intellectual but also strongly emotional responses to their thought and actions, whether of a heightened admiration or disgust. The founders are literally so present today that they even appear capable of creating their own twenty-first-century fashion: the recent term

Founders Chic[17] sounds like an echo of their eighteenth-century world in that it describes a popular trend in historiography as an aesthetic rather than a political or intellectual problem. As discussed in this book, the founding generation had similar anxieties regarding the aesthetic charms of a personalized historiography. During the first wave of revolutionary biographies two hundred years ago, they likewise feared the sensory appeal of historical writing as morally seductive and capable of doing harm to unsuspecting readers.

What can be said of historical continuities concerning the founding generation at large may be especially true in Thomas Jefferson's case. Not only was he particularly interested in the aesthetics of history, whether in his apprehensions of Federalist "Founders Chic" (as in John Marshall's *Life of George Washington*) or in the creation of his own aesthetic (as in his Republican style of dress as American president or in the architecture of Monticello). He also had, as has often been pointed out, an especially well developed talent for transcending the limitations of his historical period. In a discourse initiated by the alleged last words of his competitor-friend John Adams, Jefferson somehow managed to "survive" in the end. A self-proclaimed "forward-looking" thinker attempting to occupy a temporal middle ground between his own and later generations of Americans, he seems to have possessed secret keys to understanding a larger American identity.[18] Because of this survival in a quasi-transhistorical sphere, he has continued to provoke strong reactions, for some time to the advantage, but increasingly to the disadvantage, of his historical reputation.[19]

This eventful posthumous life at first seems strangely at odds with Jefferson's own political theories. Famously elaborating on the "principle that the earth belongs to the living and not to the dead,"[20] he was one of the inventors of an American tradition of newness whose exponents would continue to question all inclinations, as Ralph Waldo Emerson would have it, to "grope among the dry bones of the past, or put the living generation into masquerade out of its faded wardrobe."[21] Since Jefferson's concept of generational sovereignty sought to restrict the "powers" of the dead over the living, it might be concluded that he was not interested in how the chic of his faded wardrobe would be judged after his death. As with other founders, however, the contrary was the case. Jefferson was intensely interested not only in the long-term future of the American republic but also in the impression his own character would make in this future. His writings as a whole follow the pattern exemplified by the Indian chief whom he introduced into his *Notes on the State of Virginia:* like Logan's oration, Jefferson's statements can be read, to some extent, as a rhetorical effort designed to make their speaker a "monument" in American history. In a republican culture, as Logan's oration illustrates, this aim could best be achieved by conspicuous claims of personal disinterestedness, by reflections on the disappearance of one's historical period and the likely oblivion of posterity, and, ultimately, by the ideal form of republican self-effacement made possible by the death of the speaker.[22]

Despite his generational arguments, then, Jefferson did not see his own death as a fundamental rupture in, or possible end of, the development of his historical reputation. To the contrary, he tried his best to anticipate the requirements for the creation of a good posthumous impression, to the extent that the thought of posterity can be assumed to have powerfully influenced his writing. When studying his conscientiously preserved collection of papers, therefore, readers should keep in mind that it was significantly shaped by this special temporal constellation: the writer and collector's expectation that later generations of historians would read and judge it with an eye on its moral content. Jefferson displayed no exaggerated timidity in giving these historians hints concerning where he liked to see himself and his fellow republicans placed. Characterizing the framers of the federal Constitution as "an assembly of demigods" and the revolutionary period as a "Heroic age," he laid the groundwork for the American version of the ancient myth about a last Age of Heroes before the Iron Age of ordinary human history.[23]

As can be concluded from Jefferson's case, the timeless, almost mythic status that the founders have achieved in American culture was, to some degree at least, their own construction. To dissolve the spell of the magic twilight with which they have surrounded themselves is not easy. However, there are what strike me as two equally promising methods of dealing with this problem. One is altogether to deny the "founding fathers" the historical attention they sought and focus instead on other important cultural, social, and political issues in the period.[24] The other method, of course, is to continue studying them. As the impressive range and sophistication of critical Jefferson scholarship has demonstrated, the better known his writings have become, the more scholars have found to criticize in them, and the more nuanced an understanding of his period has become possible.[25] The following chapters partake in this effort to remove the founders in general and Jefferson in particular from the "Heroic age" in which they hoped eternally to dwell. The focus is on Jefferson's particular approach to temporality in order to reach a better understanding of the contingent origins of his influential interpretation of American history.

Temporality and Slavery

I was originally drawn to the topic of Jefferson's thinking about time and history for two reasons. First, the abundance, stylistic creativity, and, in part, eccentricity of his reflections on temporality may be considered interesting phenomena in their own right. Second, the topic suggested itself in a previous study of what tends to be recognized today as Jefferson's greatest moral problem, his identity as a slaveholder combined with his procrastinating attitude in the slavery question.[26] Jefferson's views on slavery can be understood as a key to his thinking generally, because slavery represented the essence of what the Enlightenment claimed to be fighting against.[27] Epitomizing radical dependence and unfreedom, slavery was the antithesis of Enlightenment–as the economic, legal, social, and political institution of chattel slavery and as a catchphrase for many other forms of oppression. From Jefferson's point of view at the end of his life, he had devoted all his time to the fight against slavery, from the American Revolution, the "Revolution of 1800," and his presidency to his final retirement, when he composed didactic writings supposed to end the enslavement of mankind in the future.

Owing to its centrality in Jefferson's thought, the topic of slavery finds its way, in different aspects, into each of the chapters that follow. In the specific question of chattel slavery, Jefferson's legacy–and in this point he may be seen to represent the founding generation at large–turned out to consist mainly in the postponement of an answer to an uncertain national future. To some extent, therefore, the problem of slavery in the American founding period can itself be viewed as a problem of timing. When Jefferson reflected on slavery and on what was becoming the question of "race,"[28] references to time played a decisive role. In writing about slavery and race, he seems to have been especially careful in crafting his temporal specifications. Nevertheless, many of his elaborate reflections on time strike modern ears as particularly unfamiliar in this context. The timing of Jefferson's views on slavery can therefore also serve here to illustrate the starting point of my study, my assumption of a partial remoteness of Jefferson's Newtonian temporality from more modern conceptions.

As the author of the Declaration of Independence and one of the largest slaveholders of his native state, Thomas Jefferson appears today the embodiment of the dialectics of the American Enlightenment: his life and thought united the expression of the rights of man in the famous lines of the Declaration and their radical negation in the institution of chattel slavery. After a war fought to end British measures to "enslave" the American colonists, Jefferson and his generation continued, in their newly founded republic, to profit from an institutionalized form of slavery that restricted the liberty and happiness of its victims to a far greater degree than did the "slavery" of taxation without representation. It was in his interaction with those outside the slave system, it has often been pointed out, that Jefferson most clearly voiced his awareness of this contradiction. As the American minister to France during the 1780s, he was especially interested in conveying a favorable image of himself and his native country to an enlightened European audience. Trying to achieve this aim by improving the articles on the United States in the *Encyclopédie méthodique,* Jefferson linked the problem of slavery to the problem of time perception: "What a stupendous, what an incomprehensible machine is man! who can endure toil, famine, stripes, imprisonment & death itself in vindication of his own liberty, and the next moment be deaf to all those motives whose power supported him thro' his trial, and inflict on his fellow men a bondage, one hour of which is fraught with more misery than ages of that which he rose in rebellion to oppose."[29]

Recognizing the morally compromised position of the American revolutionaries, Jefferson sought to explain the recent failure of the Virginian assembly to attach an emancipation clause to its new slave law. In an argument that has mostly lost its self-evidence today, he ascribed this failure, not to a lack of "virtue & firmness" of individual Virginian citizens, but to a problem of timing in the "incomprehensible" mechanism of "human nature." As he saw it, human beings were able to change their perspective on human rights from one moment to "the next moment," binding their fellow men to a system of chattel slavery, in which a single "hour" could be harder to endure than "ages" of political oppression. The problem of slavery, thus illustrated by the contrast between two modes of time perception, could only be solved by the passing of time itself: "But we must await with patience the workings of an overruling providence, & hope that that is preparing the deliverance of these, our suffering brethren. When the measure of their tears shall be full, when their groans shall have involved heaven itself in darkness, doubtless a god of justice will awaken to their distress, and by diffusing light & liberality among their oppressors, or at length by his exterminating thunder, manifest his attention to the things of this world, and that they are not left to the guidance of a blind fatality."[30]

Whereas Jefferson's recognition of the double standard of the American Revolution in the previous passage is directly accessible to today's sensibilities, his conclusion is far less so: although he realized that the continuation of American slavery was a glaring injustice that threatened to divest the American Revolution of some of its credibility, he was still able to draw attention to this fact at the same time that he openly recommended "patience" as the best possible mode of dealing with the problem. Jefferson's views differ from today's perspective on eighteenth-century American slavery, it may be concluded, not in the recognition that the institution was morally wrong and had to be abolished but in the significance he attributed to the *time* of abolition.

While the particular intensity of this passage may have been due to his collaboration with the French editor of this section of the *Encyclopédie,* the structure of the argument was characteristic of Jefferson's lifelong views on slavery. He never wavered in his conviction that slavery was an "injustice" and an "evil" that had to be abolished: it was an unpardonable "injustice" for the slaves, according to the ancient argument that slavery was a continued state of war (which he had come to recognize as an unjust war), and it was a great "evil" for the masters, as it threatened to corrupt their manners and increase the existential danger of an American civil war. However, Jefferson became equally convinced that abolition would have to wait until some more favorable moment in the future, a moment that, it turned out, did not arrive during his lifetime. In the passage quoted above, he framed the abolition of slavery in terms that made it appear desirable that the emancipation of the slaves would be gradually "prepared" rather than take place suddenly. Abolition, according to Jefferson, would consist either in the gradual enlightenment of the slaveholders or in the catastrophic event of a divinely encouraged war between masters and slaves that would end, he had reasons to fear, in the "extirpation" of the unjust masters.[31] As Jefferson's religious terminology indicates, he was able to imagine the problem of slavery essentially as a problem of God's justice: American citizens like himself could do nothing in the present but display, as did their "suffering" slaves, the virtue of "patience" and wait for a just Newtonian God either to spread Enlightenment or to intervene directly and wind up his clockwork universe when it became necessary. According to this providential argument, it was certain that the institution of slavery would be abolished in the future, but it could not be known, from a limited human perspective, when exactly this would take place or under which conditions. Arguing that the emancipation of the American slaves had to be a gradual, long-term measure, Jefferson employed a secularized reference to Christian teleology and transferred the traditional hope for "deliverance" from a future in the next world to an unknowable future in this one.

In retrospect, Jefferson's postponement of the slavery problem to an indeterminate future appears to constitute the greatest moral flaw of his personal life and possibly of the American founding period generally. Nevertheless, as the above passage shows, he was able to recommend this postponement overtly in his writings, mostly without condescending to make personal excuses or at least moderate the rather pompous style of his utterances. This conspicuous lack of apologetic efforts, combined with his unwillingness to cut down on what strikes modern ears as a disturbingly self-righteous moral rhetoric–for instance, his unashamed recommendation of "patience" until "the measure of their tears shall be full"–is especially remarkable for two reasons that concern Jefferson's qualities as a writer. First, he was known among his contemporaries for his "masterly pen."[32] Owing to his talents as an accomplished stylist, Jefferson's political success was often suspected to be the result, to a large degree, of his potentially insincere mastery of *Ars est celare artem,* his ability to gloss over logical contradictions in his political writings by his literary craftsmanship and his talent for adjusting the stylistic nuances of his letters to the positions of his correspondents. Although his neoclassical art of writing has lost much of its appeal since the Romantic idea of artistic "genius," it is still difficult, as has often been remarked by his biographers, to reconstruct Jefferson's "true" personality behind the cautiously nuanced construction of many of his writings.[33] This is the case also for a second reason (as mentioned earlier): almost all of Jefferson's preserved writings were characterized by the temporal dimension of posterity, that is by his expectation that they would be read by later generations of Americans. To some extent, his statements were habitually controlled by the internalized anticipation of a future public opinion. It was essential to Jefferson's progressive thought in general and to his argument on slavery in particular that he expected future generations to abolish slavery, because he thought they would be, in the long run, more enlightened and morally more refined than his own. Hence, when Jefferson thought about his own and his culture's place in future American historiography, he was very well aware that slavery would be regarded as a severe moral "reproach"[34] by later historians of the American Revolution and that his own contradictory role as an enlightened slaveholder could seriously impair his personal historical reputation as well. It may therefore be assumed that he was particularly sensitive to the problem of slavery in his writings and did all that was in the power of his "masterly pen" to conceal or brighten what he saw as his darker side in order to present his position on slavery in a favorable light to posterity.

The fact that Jefferson's writings on slavery have nevertheless failed so plainly to achieve this end, from the moment when they came under closer scrutiny, in the 1960s, cannot therefore be attributed to a lack either of talent or of motivation for dissimulation. Jefferson was not a particularly honest or spontaneous writer; he knew that slavery was wrong, and he anticipated the condemnation of later generations. Yet today it seems difficult to believe that his recommendation of "patience" until the "groans" of the slaves "shall have involved heaven itself in darkness" was part of a rhetorical effort to improve the image of the slaveholding United States in the eyes of an enlightened European audience, that it was even an instance of his most radical criticism of slavery and one of his clearest expressions of sympathy with the slaves themselves (whom he rarely described as "brethren" in other contexts). For Jefferson, the reference to God's justice in the slavery problem appears to have implied a radical form of criticism by which he hoped to impress his progressive French contemporaries and posterity alike, as he expected the *Encyclopédie* to "go down to late ages."[35] Today, however, these lines tend to be read merely as a hypocritical tactic of temporizing, badly disguised by a hackneyed reference to an "overruling providence" and similar factors that evidently eluded his personal responsibility.

This obvious discrepancy between Jefferson's and today's sensibilities suggests that there are points in his argument that appear morally reprehensible in today's culture but did not strike Jefferson as problematic, as he made no efforts to hide them in his otherwise cautious, posterity-oriented writings. While he took great care, for instance, to create the persona of a detached critic of slavery, his linguistic restraint seems entirely to have left him in his illustrations of the appropriate timing of antislavery reform. Since to modern ears his references to temporality sound so completely misplaced, explanations based on the supposition of Jefferson's hypocrisy as an enlightened slaveholder may describe a personal character trait, but they do not seem to go to the root of his unabashed recommendation of patience in the slavery question. In contrast to his twisted arguments and the resulting shades of gray in his writings on other matters (as in what was becoming the crucial question of race), he dared stylistically to pull out all the stops to demonstrate his views on the timing of abolition. That he tended to draw attention to the factor of time–a factor that is a central weakness of his argument from today's perspective–may therefore suggest that Jefferson's conception of temporality was a major trait distinguishing his thought from Romantic and post-Romantic thinking.

"Time will outlive this evil also"

Jefferson's argument for the abolition of slavery as a long-term measure largely reflected a consensus among European and Euro-American thinkers of his generation,[36] until it came to be criticized as "gradualism" from the late 1820s and early 1830s on by the first generation of outspoken immediatists.[37] Even arguments that explicitly abstained from voicing racist or protoracist prejudice, such as those by Jefferson's friend Condorcet,[38] tended to recommend gradual rather than immediate abolition plans, mostly based on environmentalist arguments that the slaves needed time to acquire the proper habits for citizenship. Opponents of slavery differed, however, on the temporal scope of their gradualism, as is illustrated by the various slavery reforms in the different northern American states in the late eighteenth and early nineteenth centuries.[39] Jefferson's own views on the best time for initiating a gradual abolition plan changed significantly in the course of his long life. Roughly speaking, his eighteenth-century writings tended to envision a change of the status quo of slavery happening within his own lifetime, with a heightened sense of urgency in the last decade of the century. In his nineteenth-century arguments, by contrast, the beginning of a gradualist reform receded into a more and more remote future.

The text among Jefferson's writings that came closest to suggesting a speedy beginning of gradual abolition measures in the entire United States was his draft of the Declaration of Independence, owing less to the radical implications of the phrase "all men are created equal" (whose fuller recognition would be a long time coming) than to a point in Jefferson's list of violations of American rights that was struck out by Congress. In a paragraph accusing George III of having "waged cruel war against human nature itself," Jefferson had blamed the continuation of the slave trade entirely on the British monarch: "Determined to keep open a market where MEN should be bought and sold, he has prostituted his negative for suppressing every legislative attempt to prohibit or to restrain this execrable commerce."[40] Jefferson's argument that George III had personally "deprived" American slaves of their "liberty," supposedly against the will of the American colonists, seems outrageously self-righteous at first sight. In the context of Jefferson's larger interpretation of American history, however, this passage appears in a different light. The continuation of slavery for almost another century would have been much more difficult to justify in a historical narrative of American progress if the founding document of the republic had listed the violation of "the most sacred rights of life and liberty," in the specific context of chattel slavery, among the historical "causes" for independence. In the rigid historical "chain of causes and effects" that Jefferson constructed in the Declaration of Independence, the removal of a "cause" that had "impelled" Americans to separate from the mother country–the king's determination to continue the slave trade–would have implied the elimination of all its effects upon the achievement of American independence, the elimination not only of the Atlantic slave trade itself (which was officially abolished at its earliest constitutional date during Jefferson's presidency) but also of the institution of slavery, which tended to be regarded as an effect of "this execrable commerce" in the Enlightenment. Hence, it is not surprising that the later-famous (but at the time more common and, according to John Adams, "hackneyed")[41] phrase "all men are created equal" passed the revision by Congress, while Jefferson's passage on slavery and the slave trade was struck from the final version.[42]

Jefferson never ceased to lament Congress's alterations to his Declaration draft. His increasing renown as "author" of the Declaration of Independence became a central aspect of his efforts to shape his future historical reputation: he inserted his full Declaration draft into his *Autobiography* and listed his "authorship" (rather than a less inclusive "penmanship") of the Declaration as the first of three major achievements in the epitaph he designed for his tombstone.[43] However, Jefferson's active public criticism of slavery dwindled decisively in the decades following the achievement of American independence. In the early 1780s he tried to restrict slavery in the entire Western Territory in the (eventually abortive) "Ordinance of 1784,"[44] but from then on his criticism of the institution was mostly restricted to private letters or texts intended for an enlightened European audience, such as his *Notes on the State of Virginia* or his comments quoted above on Démeunier's article in the *Encyclopédie méthodique.* In such texts, addressed to readers already critical of slavery–enlightened contemporaries and an even more enlightened posterity–Jefferson openly argued for the postponement of initiatives against slavery until he could be sure that they would achieve a political majority among Virginian citizens. He typically reasoned, as in the above-quoted comment on the *Encyclopédie méthodique,* that "the moment of doing it with success was not yet arrived" and that "an unsuccessful effort, as too often happens, would only rivet still closer the chains of bondage, and retard the moment of delivery to this oppressed description of men."[45]

The further Jefferson moved away from actively criticizing slavery, the more often he referred to questions of time and timing, while increasingly suppressing the term *slavery* and allusions to the slaves themselves. In his last letter on slavery, written a few weeks before his death in 1826, references to time entirely replaced explicit references to slavery:

> The subject of your letter of April 20, is one on which I do not permit myself to express an opinion, but when time, place, and occasion may give it some favorable effect. A good cause is often injured more by ill-timed efforts of its friends than by the arguments of its enemies. Persuasion, perseverance, and patience are the best advocates on questions depending on the will of others. The revolution in public opinion which this cause requires, is not to be expected in a day, or perhaps in an age; but time, which outlives all things, will outlive this evil also.[46]

Thus the aged Jefferson had established a necessary link between slavery and time: he had denied himself "permission" to speak about the fate of the American slaves–to the extent that he was unable to use the term *slavery* in the whole letter on this subject–by making this permission conditional on time. Jefferson described the length of his self-imposed silence and avoidance of "ill-timed" behavior as determined by a specific historical moment ("time, place, and occasion") that had not yet arrived but could be expected to arrive sooner or later. He regarded the alliterative virtues "persuasion, perseverance, and patience," each of which implied a process in time, as the best mode of dealing with slavery. As was absolutely clear by the spring of 1826, the "revolution in public opinion" that would eventually solve the problem would not occur during his own lifetime and might still require "an age." Yet according to Jefferson, it was beyond doubt that this historical moment would finally arrive, since time would in any case be able to "outlive" corrupt human institutions like slavery.

In this last personal letter on slavery as in other texts, such as the earlier passage in the *Encyclopédie méthodique,* Jefferson took great pains to describe the problem of slavery as a problem of time. It is remarkable that he did not emphasize another dimension: discussing slavery as a problem of space would seem to have suggested itself more easily, since the spatial separation and "expatriation" of the slaves were, for him, always necessary conditions for the end of American slavery. Instead, Jefferson chose to stress the links between slavery and a twofold experience of time, a powerful universal time on the one hand and the relative human experience of time on the other. This dual conception of temporality appears to have provided certainties that somehow eroded in the course of the following two centuries. From a twenty-first-century perspective, Jefferson's specifications, whether regarding the nature of time generally or the relationship between his envisioned historical moment and his conception of historical agency, seem quite enigmatic. What precisely made it so safe for him, for instance, to assume that time would end all evils, including slavery? How could he conceive of a time that would itself engage in temporal behavior and "outlive" all things? Put differently, if he considered time alive enough to "live" longer than the rest of the universe, did he also expect it to "die" one day? Why did he not refer to entities such as the republic, the nation, or "history" as outliving all evils? And if an all-surviving time was the only remedy, how exactly did Jefferson conceive of the historical "revolution" that was still necessary to effect the end of slavery? Why were time-related virtues so important in this revolution? And where did he see himself in it: was the letter a polite refusal to make a statement on slavery, or did he in fact allude to the approaching end of his self-imposed silence through the publication of his papers after his death? How, finally, is one to read this argument from the perspective of a later historical moment, when time (whatever precisely Jefferson meant by it) *has* actually "outlived" slavery as his generation knew it?

The Spirit of the Times

Such questions about Jefferson's attitude toward the end of American slavery–and thus about the meaning he gave to the American Revolution–require a close examination of his approach to time and history. In the voluminous scholarship on Jefferson's life and writings, however, his Newtonian interpretation of what he experienced as a changing "spirit of the times"[47] has not been analyzed, whether in the context of slavery or on its own. Only a few authors have discussed aspects of his temporal thinking in his personal life,[48] and the studies on Jefferson's historical thought have been more concerned with the political import of his Whiggish conception of Euro-American history[49] than with the larger philosophical and literary background of his perspective on the past. In recent years, since the liberal-republican divide in the interpretation of the revolutionary period has largely been bridged, there has been an increasing interest in Jefferson's construction of himself as a historical figure. This interest is reflected in the study of Jefferson as a didactic "historian" who linked an interest in his posthumous "fame" to the aim of promoting the universal spread of his version of republicanism.[50] While the following chapters build on these stimulating insights, they focus less on Jefferson's immediate political intentions than on the evolution of the theory of history, in particular the complex combination of ideas about change in time, that made his approach to his own position in American history possible.

There is hardly a text by Jefferson, from his private letters to public documents like the Declaration of Independence, that does not contain reflections on temporality, whether on the level of his personal life or on a collective historical level. To some extent, his preoccupation with time reflected an attitude influenced by the ancients:[51] he often sought to test his own degree of enlightenment by pondering his relation to antiquity on the one hand and to a classical "posterity" on the other. Yet Jefferson's interest in time went further than a critical emulation of his forebears from antiquity to the Renaissance. Therefore, as I see it, a "classical republican" interpretation of time alone cannot explain his temporal sensibility. Unlike Machiavelli and his contemporaries, for instance, Jefferson and his peers were able to admire Newton's science, identify with Sterne's time-conscious characters, grieve about the loss of an Ossianic past, listen to Beethoven's rhythms, or count the seconds on their watches. Their late Enlightenment experience of time was, in crucial aspects, different from that of their intellectual ancestors in previous centuries. And in equally decisive ways, it differed from later modes of time perception: as is well known, Jefferson's generation was never in a position to study train schedules, send telegrams, vulgarize Einstein or Bergson, enjoy jazz music, or chat on the Internet.

In order to appreciate Jefferson's transitional experience of change in time, I have found it helpful to understand the intellectual history of temporality–like that of other topics–as an accumulative process that relies more on addition than on subtraction. From antiquity until today, this process has engendered new emphases in the perception of time without systematically replacing older views. Thus, classical republican conceptions of time may still be felt in Jefferson's writings, but they are unable, by themselves, to account for his attitude toward change. And Newton's laws can remain useful today in the limited context of high-school physics, although they have long ceased to be the standard for more complex scientific questions.[52] Similarly, many problems in Jefferson's approach to time remain with us in a modern or postmodern world, but it is important to keep in mind that even in his most "forward-looking" moments he could never have become aware of ours.

The development of modern music, which in the Enlightenment was theorized as the prototypical temporal art, may briefly illustrate this accumulative conceptualization of time over the centuries. The early classical and classical music that Jefferson cherished was characterized by the radicalization of two contrasting developments in the perception and aesthetic expression of temporality. On the one hand, eighteenth-century musicians tried to perfect their time measurement, an effort culminating in the invention of the metronome in the early nineteenth century. (Jefferson, as it happens, was actively involved in the invention of the plexichronometer, an early version of the metronome whose mechanical measuring of beats per minute eventually made a standardized control of pace possible.)[53] Yet on the other hand, eighteenth-century composers and performers showed a growing desire to express their individual experience of time, in increasing variations of pace, rhythm, agogic accents, and so on. The ensemble of these two opposite tendencies–a greater emphasis on the quantifying and the universal aspects of time, on the one hand, and on time's qualifying and individual aspects, on the other–produced a tension that has been a source of creativity in the development of modern music. Without ever remaining the same, it has had important practical consequences from the generation of Bach's sons until today.

As I argue in the following chapters, Jefferson and his contemporaries were beginning to experience this tension–a tension between an absolute time of natural laws and a relative time of human actions and perceptions–not only in music but also in various other fields. For philosophers concerned specifically with the ontology and epistemology of time, the Newtonian assumption of differing temporalities was becoming increasingly dissatisfactory in the late eighteenth century. For the majority of Jefferson's contemporaries, however, the plural within expressions like *spirit of the times* or *course of human events* did not necessarily produce dissonance: as in their musical practice, it remained possible for them to reconcile it to an aim of universal harmony. Jefferson's dual experience of time, as has been said, was characterized by his sensitivity to both positions. In his long and eventful life, he was often brought to the verge of realizing the potential of conflict between the different "times" in his revolutionary world. As in other contexts, however, he usually remained intent on expressing what he called, in the context of the Declaration of Independence, "the harmonizing sentiments of the day."[54] Accordingly, in contrast to the contemporary avant-garde in time philosophy, he tried his best to integrate the divergent temporalities into his arguments. When their inner contradictions became evident, and his open-minded flexibility difficult to maintain, he tended to refrain from questioning further and instead increased the harmonizing efforts of his "masterly pen." As a result, his writings now offer a treasure of late Enlightenment arguments on time and history, complex and varied, as well as largely unburdened by consistent efforts to create new categories and move beyond the thinking of his period.

This book takes up Jefferson's holistic approach to time as a personal, political, literary, scientific, artistic, economic, social, and philosophical problem. The six chapters are divided into two parts. Part 1, "Time," is organized systematically by the Newtonian concepts of time that provided the groundwork for Jefferson's thinking. Part 2, "History," is structured chronologically, following the evolution of Jefferson's historical thinking within the limits of his Newtonianism. The three chapters of part 1 begin by examining the ontological and epistemological problems at the root of Jefferson's experience of the passing of time. They go on to focus on these temporalities on the level of individual experience–in the structure of Jefferson's plantation economy at Monticello and his interest in the natural sciences (chapter 1), in his opposition between "natural aristocrats" and European women, American slaves and Romantic youths (chapter 2), in his literary tastes, and in his modes of dealing with human finitude and death (chapter 3). The three chapters of part 2 analyze Jefferson's approach to change in time on the collective level of American history. They trace the evolution of his narrative of American progress from the 1770s until his death in 1826, from its theoretical foundations as he gained them in his early reading (chapter 4), through the crises in the historical thought of his later years, with political implications for his views on the French Revolution, Native American history, and the future of American slavery (chapter 5), to his efforts to come to terms with the complexity of history in his final retirement, as in his literary-historical attempts to influence the image of the American Revolution in the eyes of posterity (chapter 6).

Thus, this book places Jefferson's thinking in a unique period of intellectual history, a period, roughly speaking, between premodern and modern modes of explaining change in time. From today's perspective, the eighteenth century often comes across as an age of agelessness–as ageless as Jefferson's contemporaries looked in their artificially grayed wigs–an age characterized, more than earlier and later centuries, by a peculiarly "unhistorical" approach to historical change. Jefferson's self-construction in the realm of a mythic "Heroic age" epitomizes this approach. However, such impressions of classical timelessness resulted, not from a lack of reflection on time and history, but only from a different kind of reflection. As I shall argue, Jefferson tried to tame the contingency of historical change by organizing human concerns according to a universal Newtonian temporality, which he assumed existed independently from human perception. It was only when these attempts failed–as during the crises of his middle and later years–that he began to search for alternative modes of explaining change in time.

PART I

TIME

Thomas Jefferson's thinking was characterized by the remnants of an age-old dualism of time and eternity that had begun to disintegrate. In his eighteenth-century universe, secular time tended to expand, at the expense of eternity, both into the past and into the future: on the one hand, the world became older as the date of the Creation was located in an increasingly remote past; on the other hand, the secular future of mankind became the subject of a growing number of literary, philosophical, and political reflections. In the process, the contrast between time and a "timeless" eternity tended to fade, taking the form of two different secular concepts of time. These two "times" in the transitional thinking of Jefferson and his contemporaries–a universal time of natural laws and a particular time of human perceptions and actions–are the subject of the following three chapters.

The secularization, or "temporalization,"[1] of eternity was condensed in the conception of an absolute time existing in its own right: Isaac Newton's "absolute, true and mathematical time," which, according to his much-criticized definition, "of itself, and from its own nature, flows equably without relation to anything external."[2] This absolute time was assumed to be universal (unrelated to sensible objects and their position in space) and homogeneous (flowing at a constant rate). It existed in "its own nature," as a reality independent from human perception. Newtonian absolute time can be understood, to a certain extent, as a secularization of traditional concepts of eternity. Positively defined as an everlasting duration, absolute time, in the cautious arguments of Newton and his students, was the emanation of a God who had ceased to reside "above" or "outside" time. God's existence was no longer timelessly eternal, but became sempiternal, always present in the infinite, universal, and equable flow of absolute time.[3]

Newton assumed that human time perception was approximative. He defined two different concepts of time: apart from absolute time, there was also a "relative, apparent and common time." If absolute time was the secularization of eternity, this relative time was a secularization of traditional concepts of a "merely" human time, a temporality dominated by fortune and chance, by chaotic human actions and limited human perceptions. Unlike Newton's mathematical absolute time, relative time was measurable by the senses. In his words, relative time was "some sensible and external (whether accurate or unequable) measure of duration by the means of motion, which is commonly used instead of true time; such as an hour, a day, a month, a year."[4] For Newton, relative time was useful in everyday concerns but should not be the standard in natural philosophy. In retrospect, that this empirical thinker held such an attitude might seem surprising, but Newton was still caught up in a complex theological argument. Although his intellectual heirs among later natural scientists would separate absolute time from theological questions,[5] Newton considered the knowledge of his quasi-divine absolute time to be mediate and deductive rather than inductive: natural philosophers were supposed (rationally) to deduce the existence of absolute time from absolute movement, which in turn they had to deduce from their comparisons of the relative movements they (sensorially) perceived. Thus, their observations could approach, but never quite reach, the measure of absolute time.

The slowly dissolving Newtonian dualism of absolute and relative time and the approximative empiricism that resulted from it provided the structure of many eighteenth-century reflections on temporality. Newton's attempts to come to terms with the aporia of objective time measurement created new tensions that eventually inspired Immanuel Kant's *Critique of Pure Reason* (1781, 1787). Kant's arguments put an end, in the long run, to the approximative character of the Enlightenment by uniting crucial aspects of Newton's two concepts within a single definition of time as a pure form of intuition that was at once empirically real and transcendentally ideal. Jefferson, of course, never studied Kant's *Critique* and did not think of allowing time the status of a formal a priori condition of all appearances.[6] While his reflections on temporality were already characterized by the problems leading up to Kant's transcendental idealism, they ultimately failed to move beyond the pattern of absolute and relative time. In his version of a Newtonian empiricism, Jefferson continued to assume the existence of an absolute and homogeneous "flow" of time according to universal natural laws, while simultaneously understanding human time perception as relative, heterogeneous, and at best leading to an approximative estimate of absolute time.

To be sure, the details of the ontological aspects of Newton's argument did not play a large role in the temporal reflections of a man like Jefferson, who typically professed an intense dislike of questions he considered "too metaphysical."[7] Much as he admired the figure of Newton, another author appealed more immediately to his taste in this context: John Locke's discussion of time in *An Essay Concerning Human Understanding* (1690) had a more direct influence on Jefferson's writings. The *Essay* mirrored the ambivalence of Newton's temporal thought but dealt with the topic on another, programmatically less elitist level. Locke largely took Newton's metaphysical arguments for granted but tried to circumvent the aporia of objective time measurement by approaching the "very abstruse" problem of time from the viewpoint of epistemology:

> The answer of a great man, to one who asked what time was, *Si non rogas intelligo* (which amounts to this: The more I set myself to think of it, the less I understand it), might perhaps persuade one that *time*, which reveals all other things, is itself not to be discovered. *Duration, time,* and *eternity* are, not without reason, thought to have something very abstruse in their nature. But however remote these may seem from our comprehension, yet if we trace them right to their originals . . . we shall find that the *idea* of eternity itself is derived from the same common original with the rest of our *ideas.*[8]

The contradictory implications of Locke's *Essay* would occupy the discussion of time and history that became relevant for Jefferson's thinking. As described in the following chapters, Jefferson reflected on Locke's epistemological arguments directly and also found them discussed in his favorite novel, Laurence Sterne's *Life and Opinions of Tristram Shandy* (1759-67). In the above extract, Locke's position on time philosophy was condensed in his half-satirical translation of Augustine's famous expression of bewilderment at the problem of time.[9] In achieving an ironic distance from Augustine's position, yet quoting him in the beginning of the chapter as the only "great man," Locke illustrated the ambivalent attitude of his *Essay* toward Augustine's influential explanation of the subjectivity of time as a *distentio animi.* On the one hand, Locke's empirical concepts of time and memory, tacitly based on the existence of Newton's absolute time, differed sharply from Augustine's timeless eternity and his active *memoria interior.* On the other hand, however, Locke's thinking also implied a continuation of the subjectivist tradition initiated by Augustine's *Confessiones,* since it likewise focused on time perception as a subjective process within the human mind. Locke sought to demonstrate that all ideas about time, timelessness, or everlasting duration–even "the idea of eternity itself"–could be traced back to the same roots as all other ideas, namely, "sensation" and "reflection." Anticipating Kant's secular subjectivism in this respect, the perception of time was intertwined, for Locke, with the process of reflection: it was essentially tied, not to the sensory perception of external motion, but to the internal "reflection on the train of *ideas* which we find to appear one after another in our own minds."[10] Thus, Locke saw "the measure and standard" of time perception in "the constant and regular succession of ideas" within human understanding.[11]

Although less explicitly than Newton's definitions of absolute and relative time, Locke's internal time measurement by the "constant and regular succession of ideas" also allowed for two different concepts, a subjective and an objective concept of time. Without the latter, after all, it would not have been possible to assume that the internal "succession of ideas" succeeded at a pace that was "constant and regular."[12] Thus, Locke's conception of the human mind as its own little watch was beginning to undermine Newton's absolute time. Yet time and again, Locke's argument also fell back into a Newtonian universe,[13] continuing to imply the existence of an absolute time, imagined as a perfect clock that provided the standard for, but always remained partially hidden from, a limited human perception.

Like many of his contemporaries, Jefferson followed both Newton and Locke in his inability (or unwillingness) to recognize the full potential of conflict between absolute and relative time. Although he experienced situations in which their incongruity became evident, he tried his best to allow them equal places in his arguments, as part of a worldview that accepted human limitations and sought to reconcile possible antinomies in the ideal of a golden mean. A proudly eclectic thinker, Jefferson cannot be expected to have written a systematic tract, least of all on the "very abstruse" subject of temporality. Nevertheless, almost all of his writings contain reflections on time, in some cases in such prominent positions and with such frequency that they suggest a personal preoccupation with the subject. The flexible but in the long run unsatisfactory dualism in Newtonian time philosophy pervaded Enlightenment culture, and Jefferson, perhaps owing to a personal penchant for musing on his own place in time and history, appears to have been particularly responsive to the subject. In character with his professed aversion to metaphysics and to philosophical systems, one of Jefferson's most systematic discussions of the two temporalities took place in a half-serious albeit intensely personal context: in the "Head and Heart" letter that he wrote with his left hand to the Anglo-Italian painter Maria Cosway.[14]

This longest of Jefferson's Sterne-inspired love letters may serve here to illustrate, briefly, the dualistic nature of his conception of time. In this text, he presented his head and his heart in the middle of a conversation about the questions how, among other things, to measure time, to structure one's life, and to evaluate past experience. The assumption of differing temporalities structured both the form and the content of the letter. In the first place, it enabled Jefferson to create a tension between the imagined time of the dialogue, on the evening after Mrs. Cosway's departure, and the actual time of composing the letter: forced to write "slowly and awkwardly" with his left hand, as he had recently broken his right wrist, Jefferson probably needed several days to accomplish this task.[15] The entire composition can thus be understood as a Shandean play with two different temporalities, an external calendar time and the time of Jefferson's internal conversation. The first temporality found a spokesman in Jefferson's Head, the second in Jefferson's Heart.

It is ultimately useless to ask how much of this love letter, written in a world where public and private were not clearly distinguished, was meant as a private expression of Jefferson's emotions and how much was intended as a public display of his conversational sophistication. (Since it was preserved in print copy and would be among his earliest published letters, it is not unlikely that it was written for the edification of posterity as well.) Accordingly, the Head's time was not entirely public, nor was the Heart's time exclusively private. Both temporalities were concerned with questions of epistemology, practical morality, and historical consciousness that involved an individual and a collective, a personal and a historical level of time perception.[16] In this comprehensive framework, the Head scolds the Heart for being "overwhelmed with grief" at Mrs. Cosway's departure for London. As a consolation, the Head offers "the inestimable value of intellectual pleasures," which are "always leading us to something new." Along the time line of Newtonian natural philosophy, the Head focuses on the aim of prediction and the search for ever-"new" observations. As a secularization of divine foreknowledge, its foresight also has a moral dimension. Thus, the Head admonishes the Heart to "look forward" and anticipate the consequences of its actions:

> This is not a world to live at random as you do. To avoid those eternal distresses, to which you are forever exposing us, you must learn to look forward before you take a step which may interest our peace. Everything in this world is a matter of calculation. . . . We ride serene & sublime above the concerns of this mortal world, contemplating truth & nature, matter & motion, the laws which bind up their existence, & that eternal being who made & bound them by those laws. . . . Let the bustle & tumult of society to those who have not talents to occupy themselves without them.[17]

If this is read as a reflection on time, Jefferson's Head seeks to approach Newton's "absolute, true and mathematical time" in all three characteristics: Since it regards everything as "a matter of calculation," its time perception has a quantifying, "mathematical" character. Moreover, by contemplating "truth & nature," the Head tries to gain the knowledge of Newton's "true" time. And finally, the Head emphasizes the quasi-divine attributes of "absolute" time when it attempts to "ride serene & sublime above the concerns of this mortal world," trying to emulate God's foresight by avoiding a "random" human life.

Jefferson's Heart, on the other hand, bluntly refuses to ride "above the concerns of this mortal world," greatly enjoys "the bustle and tumult of society," and wishes nothing more than to live the past love story with Maria Cosway over and over again, at least on the level of narrative repetition. Unlike the Head, it accepts its human temporality. Criticized for its inability to "look forward," the Heart defends its lack of future-orientation by a reflection on the inevitable vanity of human life: "We have no rose without it's thorn; no pleasure without alloy. . . . It is the condition annexed to all our pleasures, not by us who receive, but by him who gives them." In its present-centeredness, the Heart's "rose" might be seen as a precursor to the transcendental roses under Ralph Waldo Emerson's window in *Self-Reliance* (1841). Unlike Emerson's roses, however, Jefferson's Heart does not live exclusively in an everlasting present, "above time."[18] Instead, it is oriented toward a present and past that have a specific pace of their own. Remembering the beauties of a day trip with Maria Cosway, Jefferson's Heart constructs its own, particularly human time by attributing to each recollection a different pace of temporal change: "The wheels of time moved on with a rapidity of which those of our carriage gave but a faint idea, and yet in the evening, when one took a retrospect of the day, what a mass of happiness had we travelled over!"[19]

Unlike the Head's approximation to an "equably" flowing time, the Heart's time sometimes flows faster, sometimes more slowly. Unlike the Head's, its passing cannot be deduced from an absolute motion derived from the comparison of different relative motions in space. By stressing that the "wheels of the carriage" gave "only a faint idea" of the "wheels of time," Jefferson's Heart acknowledges the aporia of objective time measurement. Emphasizing the insufficiency of natural laws to do justice to human time perception, it begins to emancipate itself from the Head's regularity. For the Heart, the real standard by which to evaluate a past experience is not its quantity but its quality, not a uniform motion in space but the great "happiness" Jefferson had experienced.

The Head's rational time and the Heart's sentimental time are also relevant for Jefferson's historical consciousness. The Head judges the "history"[20] of his past flirtation with Maria Cosway only according to its didactic value for his future, complaining to the Heart: "I reminded you of the follies of the first day intending to deduce from thence some useful lessons for you, but instead of listening to these, you kindle at the recollection, you retrace the whole series with a fondness which shews you want nothing but an opportunity to act it over again."[21] Whereas the Head's conceptions of time and history are linear and forward oriented, those of the Heart tend to be cyclical and backward oriented. While the Head sees Jefferson's past mostly as a collection of moral "lessons" of a didactic philosophical history, his Heart develops an interest in the past as such, fondly recollecting and retracing the past events for their own sake. In its approximation to the absolute time of natural philosophy, the Head tries to turn Jefferson's past into the typical Enlightenment narrative of his past "follies," useful only as examples of what to avoid in the future.[22] However, it is not successful in educating the Heart, which clings to its qualifying approach to the past, a past it remembers mainly for its unique emotive appeal.

After the transcendental turn in time philosophy, nineteenth-century thinkers would be under greater pressure to replace the "times" of Jefferson's approximative empiricism by a single concept of temporality. While natural scientists, for instance, would self-confidently go on to measure a universal and homogeneous time (at least until the next turn of the century), romantic artists would become more radical in emphasizing individual differences in the perception of time. Jefferson's thinking went in the direction of these modern modes of differentiation but eventually stopped short of them. From the perspective of a more fully secularized time consciousness, therefore, his plural temporality cannot wholly be grasped intuitively but requires a careful reconstruction. For this purpose, the time of Jefferson's Head, deducing a Newtonian absolute time, is labeled *rational time* in this book, while *sentimental time* designates the Heart's perception of a Newtonian relative time. These simplifying labels are not philosophically accurate terms, whether in the terminology of enlightened or of today's discussion of time. Nor are they meant to indicate that Jefferson's rational time was rational by a universal standard (indeed, he sometimes referred to it in contexts that were quite irrational by most standards) or that his sentimental time was tied to the sentimental cultivation of civilized emotions (to the contrary, it became most intense in his views on the revolutionary bloodbath in Europe). It is in their very inclusiveness, however, that the labels *rational time* and *sentimental time* promise to approach the capacious categories in the flexible mind of Thomas Jefferson. His reflections on time found one of their most systematic expressions in the humorous voice of a love letter, yet–as contended in the following chapters–this did not keep them from playing a decisive role in both his personal life and his historical imagination.

Chapter 1

Rational Time

When Jefferson's preoccupation with time is mentioned in scholarly discussions, it is usually associated with a linear, quantifying concept of time, a time that steadily progresses into the future. It is identified with the Jefferson who loved to collect clocks and watches, who kept minute records of the weather, his expenses, or even the vegetables on his table, and who generally liked "the dreams of the future better than the history of the past."[1] This penchant for measuring, recording, and controlling the flow of time in the present in order to achieve happiness in a secular future is usually regarded as the typically "enlightened," middle-class approach to time–the rational time of this study.

The label *rational time* refers to Jefferson's approximative perception of a Newtonian absolute time that had become the standard in natural philosophy. Approaching an everlasting duration, his rational time retained several characteristics from prior notions of eternity. Apart from its standard of infinity, universality, and homogeneity, rational time was distinguished by a secularized version of divine foreknowledge. For Jefferson, this secular human foresight was part of the faculty of reason, connected, in the tradition of ancient writers, to a socially privileged position. While it mostly failed to order his finances, rational foresight played an important role in the ideal life of his "natural aristocracy"[2] and in his studies of "nature" in general. Since Jefferson's conception of nature included mechanical, organic, and human nature alike, the homogeneity of rational time may be responsible for the impression of a peculiar impersonality in his reflections on human concerns: Jefferson could be rather inflexible in imposing the linear structure of rational time on his own life as well as on the lives of others. Habitual self-control and the "saving" of time in the present, he assumed, allowed human beings to "improve the future time." For him, a secular future thus often took the place of the afterlife as an incentive for good behavior. As argued in this chapter, the secularized teleology of rational time rigidly organized his own day schedules as well as those he designed for his daughters, nephews, and grandchildren.

In eighteenth-century literature, it is perhaps the title character of Daniel Defoe's *Robinson Crusoe* (1719) who best exemplifies the conception of rational time.[3] Afraid to lose, in his isolation from society, his "Reckoning of Time for want of Books and Pen and Ink," Robinson Crusoe erects a wooden cross on the shore. Marking the date of his shipwreck and each subsequent day by a notch in the wood, he keeps a "Kalandar, or weekly, monthly, and yearly reckoning of Time." On his island, of course, this calendar does not fulfill its usual function of coordinating social activities. It serves no other purpose than to provide Robinson with a visible sign of his own civilized–that is, Christian, middle-class, and time-conscious–identity. His habitual measurement of time culminates in a "Journal of every Day's Employment," which meticulously records weather conditions and allows him to plan his agricultural activities. Robinson also designs a day schedule, containing the crucial division of work and leisure time: "This morning I began to order my times of Work, of going out with my Gun, time of sleep, and time of Diversion."[4] When he finally meets another human being, the social significance of his mastery of time becomes apparent: "In a little Time I began to speak to him, and teach him to speak to me; and first, I made him know his Name should be *Friday,* which was the Day I sav'd his Life; I call'd him so for the Memory of the Time; I likewise taught him to say *Master,* and then let him know, that was to be my Name."[5]

As the name Friday indicates, Robinson establishes his authority as master also by literally imposing the temporal order of the biblical week on his servant or slave, whose life he has spared or saved. He thus reiterates a connection between time perception and slavery going back as far as Hesiod's *Works and Days,* in which the master is placed between Zeus (son of Cronos) and his slaves, distinguished from the former by an inferior knowledge, and from the latter by a superior knowledge, of time.[6] In a hierarchy still basically existent at Jefferson's Monticello, the time-conscious slaveholder stood between his slaves, on the one hand, and Newton's Great Clockmaker, on the other. This chapter traces the characteristic aspects of rational time in Jefferson's thought: its mechanical measurement, its religious origins, its secular future-orientation, its internalization by way of journals and records, and its beginning division between work and leisure time. Since Jefferson, like Defoe's Robinson, used this temporality as a means of establishing his authority as master, the chapter concludes with his views on the social significance of rational foresight.

The Great Clockmaker

Jefferson's rational approach to time strongly resembles the temporal paradigm personified by the character of Robinson Crusoe. Because of Jefferson's "almost frantic" concern with time control, one scholar has gone so far as to attest that Jefferson had a "fixation" on (rational) time.[7] A striking result of this fixation can still be admired at Monticello and the University of Virginia: the vast array of clocks and watches that he bought and helped design over the years. In his concern with time measurement, however, Jefferson was hardly alone in a period that experienced a boom in the demand for mechanical timepieces. Produced increasingly faster and cheaper (the clock-making industry itself being an early example of a time-saving division of labor),[8] clocks and watches were becoming affordable possessions for larger parts of eighteenth-century society, often functioning as important markers of social status. Reaching as far as the imperial court in Beijing, the craze for timepieces can almost be described as an early globalized fashion.[9] The groundbreaking technical innovation in time measurement had been Christian Huygens's introduction of the pendulum as part of the clockwork in 1657, making a time standard in seconds practicable. As Jefferson argued in his *Plan for Establishing Uniformity in the Coinage, Weights, and Measures of the United States* (1790), the "seconds of mean time" measured by the pendulum were best suited for a universal standard also of spatial measurements (a standard that, in contrast to the centralized French regulations of the 1790s, had the benefit of being globally applicable). Jefferson remained aware, to be sure, that "even a pendulum is not without its uncertainties." In a world whose matter furnished, in its extension, "nothing invariable," the mean time measured by clocks came closest to establishing a uniform standard but ultimately also remained a mere approximation.[10]

Making this general concession, Jefferson shared the ambition of many contemporaries further to improve the accuracy of the clockwork.[11] Its increasing precision was relevant in many fields. As president, Jefferson oversaw the United States Coast Survey, which used chronometers to chart

American coastlines and inland points for the use of mapmakers, sea captains, and settlers.[12] The precision of time technology, of course, was generally crucial for ship navigation and thus for commerce and the development of colonial expansion. In this sense, the exploration of the New World was inherently structured by a rational time consciousness that relied on the measurement of movement in space. On the individual level, the proliferation of increasingly precise timepieces was consequential as well. When church or city clocks were supplemented by household clocks and pocket watches, the nature of everyday time perception underwent lasting changes. As the frequency of casual references to minutes and seconds by eighteenth-century writers suggests, time was perceived in smaller units by increasing numbers of people. The transition from a mostly passive acoustic perception of time at fixed intervals (as by the sound of church bells) to an active visual perception of time at any moment (as on the face of one's pocket watch) contributed to a greater sense of the rationality of a continuous, regularly progressing clock time.[13] Thus, the growing technical refinement of the clockwork was both the cause and the effect of a fundamental transformation in the consciousness of time. Its multilayered significance made the clockwork the epitome of a Newtonian universe ordered by a divine Clockmaker. After what has been described as Newton's "Pyrrhic victory"–the long-term disappearance of his Clockmaker due to the practice of his physics[14]–the clockwork itself would become a summary symbol for processes of rationalization, mechanization, industrialization, and modernization.[15]

Jefferson's attachment to the symbol of a Newtonian universe was so strong that there was scarcely a room at Monticello–from his bedroom to the kitchen–without a timepiece. In this respect curiously similar, for instance, to George III,[16] Jefferson liked to overlook the construction and design of his clocks and took great care in the selection of gold and silver pocket watches for himself, his friends, and family members.[17] Besides, he was interested in various other gadgets connected to time measurement, such as the traditional sundial, an early metronome, and Rittenhouse's orrery. Jefferson praised the orrery enthusiastically in the terms of his approximative Newtonianism: as he saw it, Rittenhouse had "not indeed made a world; but he has by imitation approached nearer its Maker than any man who has lived from the creation to this day."[18] Jefferson's most famous timepiece, the Great Clock, can still be admired today: it overlooks the entrance hall in the great house at Monticello.[19] Made according to his specifications

FIGURE 1. The Great Clock in the Entrance Hall at Monticello, 1792–93. (Monticello/Thomas Jefferson Foundation, Inc., Photograph by Carol Highsmith)

in 1792–93, this clock is an interesting object not only for its craftsmanship but also as a symbol of Jefferson's philosophy of time (figure 1).

A two-faced "cannonball clock" over the entrance door, readable from both inside and outside the house, the Great Clock created a complex experience of time–visual, aural, and performative. First of all, time was visualized in different degrees of exactness. The outside view of the Great Clock gave information only of the present hour, following Jefferson's wish: "There need be no minute hand as the hour figures will be 6 I. [Inches] apart."[20] The inside view of the clock, by contrast, specified time in different ways: Roman numerals for hours, Arabic numerals for minutes, and an extra dial for seconds. Similarly to a kitchen clock that Jefferson purchased at about the same time, the Great Clock also made the day of the week visible in the entrance hall.[21] Its seven-day calendar worked by the descent of fourteen cannonball weights from the clock into the basement through openings in the floor. The two sets of iron weights, which provided the motive power for the clock, were strung on ropes and descended in the corners of the entrance front, where the days of the week were marked at equal intervals down the side walls. Thus, the current time of the week (the day and approximately, once more, the hour of the day) was indicated by the position of the weights. However, this visualization was not complete: because the ceiling of the room was not high enough, visitors to the entrance hall could see all weekdays except Saturday, which had to be inscribed on the wall of the basement below.

In addition to making time visible, Jefferson sought to render it audible as well. He ordered a bell for the clock to be installed on the top of the house, to be "heard all over my farm."[22] Apparently, he did not want to do without the traditional device of a bell sounding the hours, but hoped to regulate the time perception of masters, slaves, and employees alike. Late in life, in his instructions for the central clock of the University of Virginia, he emphasized the importance of a bell "to be heard 1½ miles under any circumstances of weather."[23] Jefferson seemed to lack confidence in the general disposition of his contemporaries–field slaves, Charlottesville citizens, or the future "natural aristocracy" at the state university–toward regular time measurement, feeling the need to subject them all to a passive, acoustic perception of clock time.

Finally, the Great Clock created an experience of time on a third level. Like the other seven-day clocks in the house, the entrance clock had to be wound each Sunday morning.[24] According to the Monticello slave Isaac Jefferson, it was the habit of the master of the house to perform this task himself, at least for the kitchen clock.[25] That Jefferson personally performed what from today's perspective might seem a rather trivial task is interesting. It obviously cannot be explained by a lack of servants or slaves at Monticello. Jefferson's going about the house to "wind up the clock" each Sunday appears to have been a behavior connected to the responsibility of the head of the family, a habit so common that it was already ridiculed, as in the caricature of the pedantic "philosopher" Walter Shandy in Sterne's *Tristram Shandy.*[26] Thus, the passing of time was illustrated not only by the mechanics of the clock but also by a performance of the *paterfamilias* that emphasized the centrality of clock time in the functioning of the household. The master's ritual act of winding seven-day clocks helped bind the extended family of the plantation to the rhythm of the biblical seven-day week.

Interpreted according to its philosophical significance, Jefferson's Great

Clock was a materialization of his Newtonian approach to time, an approach that posited the existence of a secularized eternity–an "absolute, true and mathematical time"–but emphasized basic human limitations in its perception. Dominating the main entrance and ideally heard "all over" the farm, the near-eternal time of the Great Clock occupied an elevated position in the microcosm of Monticello. Unlike the cross-shaped representation of rational time in *Robinson Crusoe,* Jefferson's main timekeeper had no explicit Christian symbolism. However, its secularization was not radicalized: the Great Clock always referred to a temporality outside the human mind, demonstrating that time perception was essentially mediate and relative to the perceiver's position in space. In other words, the Great Clock openly displayed clock time as a man-made approximation to a "true" time that went beyond its limitations.[27] Hence, despite the Great Clock's technical refinements, Saturdays, unlike the other days of the week, could not be "seen" in the entrance hall, and minutes, unlike hours, could not be "heard" in the garden. To pursue this symbolic reading further, especially the invisibility and empirical nonexistence in Jefferson's entrance hall of Saturday–the day of the Creation's completion–can be understood to symbolize an enlightened recognition of fundamental human limitations in the knowledge of time. In an age before a global or even a national standard time, this readiness to accept human relativity can be explained partly by technical problems. However, it also pertained to a mentality that, despite its appreciation of scientific innovations and the pursuit of an ever-greater exactness, considered absolute time to be the emanation of a benevolent creator. In the Newtonian world epitomized by the Great Clock, man-measured time was understood as the mere approximation of an absolute time that was ultimately beyond human reach. Thus, the Great Clockmaker continued to hold his sway over Monticello, at least through his representative, the *paterfamilias* taking charge of the clocks in his small universe.

Seen in this light, the rational clock time of Jefferson's culture has to be understood not as a sharp contrast to but rather as a continuation and modification of the linear time of the Judeo-Christian tradition.[28] Jefferson's Great Clock did not express a drastic rupture with biblical conceptions but maintained conventional ideas about time and eternity in a changing frame of reference. With each new winding of the Great Clock on Sunday morning, the temporal world of Monticello was symbolically re-created as part of a deistic interpretation of Christian teleology, an interpretation that relocated the biblical institution of the seven-day week in a secular human environment. While the prominent position of Sunday-centered clocks at

Monticello can be seen as a continued expression of a Christian (according to Max Weber, especially Protestant)[29] feeling of responsibility for the use of God-given time, this responsibility was slowly becoming independent from its religious context. In contrast to the canonical hours of monasteries or the bells of church clocks, the temporal units of the Great Clock were displayed for their own sake, with no fixed meaning or action tied to them.[30] As symbolized by the Egyptian style Jefferson used for the design of the clock at the foot of his bed,[31] one hour of this secularized eternity was by definition as important as another, and it could be employed for very different purposes. Part of a beginning modern worldview,[32] this stress on the quantification rather than the qualification of time allowed more freedom for the self-reliant exercise of reason and foresight (at least for those who could see the face of the Great Clock from the inside of the great house). Time understood as something to be measured and controlled became a commodity, and the striving for a more exact time measurement could become a crucial precondition for the development of both modern economics and the natural sciences.[33]

Interpreted as a philosophical statement in a period of transition, then, Jefferson's Great Clock was double-faced not only literally but also in a figurative sense. One face looked back to a premodern world that was "merely" temporal, where all human endeavors centered on a telos that was to be found "outside" human time perception. The other face looked forward to a modern world that could only be perceived temporally. Jefferson's rational time was situated between the two faces of the clock, in the clockwork universe where a Great Clockmaker still smiled down benignly at the attempt of man to become the master of his time.

Meteorology and Money

"Mr. Jefferson rises in the morning as soon as he can see the hands of his clock, which is directly opposite his bed, and examines his thermometer immediately, as he keeps a regular meteorological diary." This 1824 description of Jefferson's temporal discipline by Daniel Webster–surprisingly intimate, coming as it did from a man who was only an occasional visitor to Monticello–is corroborated by many accounts of people closer to him. According to the Monticello overseer, Edmund Bacon, Jefferson was "always busy," so intent on saving daylight that "the sun never found him in bed." Thomas Jefferson Randolph described his grandfather as "a miser of his time," with "regular and systematic" habits.[34] Jefferson himself often referred to the rigid routine of his days.[35] His most conspicuous way of imposing a temporal order on life was his habit of keeping records–of the weather, his expenses, his legal practice, his correspondence, his stocks of wine, his vegetable plantings. In these records he conscientiously made references to time, recording the dates of spending or receiving money, of the births and deaths of slaves, of the first bloom of flowers, and so on. Jefferson's garden and farm books and especially his surviving account books–or memorandum books, as he called them–testify to the daily discipline of keeping "memoranda" at least from 1767 till a few weeks before his death in 1826.[36] He often arranged his information in charts and lists, and he developed his own abbreviation system in order to document all his data in as little time and space as possible. This rationalization gives a scientific hue to his recordings, as it gives the appearance of a systematic and objective text. However, there were two divergent modes in which rational time organized the memorandum books: the role of temporal information in Jefferson's weather charts differed substantially from its role in his financial records.

The meteorological notations in the memorandum books and elsewhere anticipated, to some extent, modern scientific standards. Jefferson has been described as a pioneer in meteorology, as "the best-informed American of his day . . . far in advance of the practices of his time."[37] Over the course of many years, he took daily notes of the weather conditions, even when he was quite busy in politics. On July 4, 1776, for instance, Jefferson had just started a new weather record and made four entries, the temperatures in Philadelphia at 6:00 and 9:00 a.m. and 1:00 and 9:00 p.m.[38] He later developed a method of measuring the temperature (to which he sometimes added air pressure, humidity, and the amount of precipitation) at what he considered to be the coldest and warmest times each day, sunrise and 4:00 p.m. From these numbers, as he explained in *Notes on the State of Virginia,* he calculated monthly average temperatures.[39] At a time when thermometers, barometers, hygrometers, and the like were still rare in America, he tried to enlist friends and family in his meteorological research.[40] His comprehensive efforts aimed at a comparative view of the American climate and a better understanding of the climate change he described in his *Notes* (a global warming known today as part of the eighteenth-century decline of the Little Ice Age). To achieve this goal, both synchronic and diachronic precision were crucial; Jefferson had to measure the temperatures at exactly the same time each year and then compare the results of different years over a longer period.

Thus, Jefferson's interest in different American *tempora* was inseparable from his attention to *tempus* in general. First and foremost, of course, this habitual attention was connected to the concerns of a Virginian planter. As exemplified by the role of meteorology in Robinson Crusoe's diary, New World agriculture was especially dependent on a rational analysis of the meteorological conditions, because frontier farmers, unlike farmers in Europe, had no handed-down wisdom about the local soil and climate to rely on. In Jefferson's case, the dominant agricultural function of meteorology even influenced his choice of paper: especially the earlier memorandum books were actually almanacs, transformed into writing paper for his diverse memoranda.[41] Hence, Jefferson's notations were literally written into the eighteenth-century genre that most conspicuously linked time to the cultivation of nature.

Yet the rational concept of time underlying his meteorological studies also had another context: meteorology exemplifies what has been described as "the very essence of scientific culture," the aim of prediction in the natural sciences.[42] Jefferson's study of American and European weather conditions was basically "scientific," practicing temporal discipline in the present in order to predict and control a secular future. As he saw it, this scientific method had to be imparted to the next generation, even to his eleven-year-old daughter Mary, whom he admonished to record the first appearances of birds, fruits, and vegetables each spring.[43] Meteorology occupied an important position in Jefferson's mind, as indicated by the place of the weather observations in *Notes on Virginia.* For him, information on Virginia's weather answered sufficiently to the extensive frame of query 7, "A notice of all what can increase the progress of human knowledge?" As he put it, under "the latitude of this query, I will presume it not improper nor unacceptable to furnish some data for estimating the climate of Virginia."[44] From today's perspective, it might appear somewhat exaggerated to consider a study of the local weather as representative of "all what can increase the progress of human knowledge." Yet in Jefferson's intellectual culture, it made sense if the "climate" of a country was seen as constitutive of the character, manners, and government of its inhabitants. He concurred with a Montesquieuean concept of climate when he claimed "that an observing traveller, without the aid of the quadrant may always know his latitude by the character of the people among whom he finds himself."[45] While Jefferson did not, of course, follow Montesquieu completely in the political implications of his climatic relativism, it remained plausible for him to think of weather observations as furthering a comprehensive scientific, moral, and political progress.

Jefferson's idea of the "progress of human knowledge" was thus structured by a concept of time imagined as continuous, homogeneous, and irreversible, a straight time line directed toward a secular future. To some extent, then, he can be understood to have anticipated a nineteenth-century scientific ethos. His cash accounts, on the other hand, reveal how rational time could function in a mainly premodern context as well. In the memorandum books, he meticulously kept track of each penny he earned, spent, or owed. Unlike in his meteorological notations, however, he mostly refrained from calculating total or average amounts. And without double-entry form, he could never be sure about the actual state of his finances and was often surprised by his towering debts.[46] The precision that characterized the references to time in his financial records therefore gave him only an illusion of control in a modern sense. Jefferson failed to anticipate long-term effects of his economic and financial decisions, as he seldom used his temporal information systematically to calculate the changing productivity and market values of his plantations, of real estate and slaves.

In a traditional agricultural slave economy, this neglect can partly be explained by structural factors. Subjectively, planters like Jefferson often seemed to lack an economic alternative that would motivate this type of calculation. After a bad harvest or in times of economic crisis, they could feel at the mercy of the credit system as much as of the weather: when their plantations turned out to be unprofitable, it could still be dangerous for planters to sell large portions of their only capital–land and slaves–at a low price.[47] Moreover, some slaveholders claimed, as Jefferson did, to have moral scruples about selling their slaves.[48] The beginnings in classical economic theory of using temporal data to calculate the productivity of labor were, of course, more relevant in the case of free than of slave labor. If Jefferson tried his hand at calculating the productivity of labor on his plantations, he did so only on occasion, as when hiring or leasing slaves or in the nonagricultural contexts of building roads or manufacturing nails. In these transitional situations, he still liked to use, besides references to hours, the premodern unit "days work."[49] His reluctance to employ rational time systematically for a comprehensive analysis of his plantation economy may be especially plausible if one takes into account his personal aversion to contemplating the exact nature of the labor on which his livelihood depended.

In this context, it is important to note that Jefferson liked to distinguish himself from slaveholders who were motivated by "sordid avarice only,"[50] a distinction that he grounded in a negative concept of financial (self-) "interest."[51] Hence, his many efforts to maximize the profits of his plantations notwithstanding, he sometimes seemed almost proud *not* to be successful financially. Financial failure, devastating as it must have appeared in the second half of his life, at least conformed to Jefferson's ideal motivation of "disinterestedness," "generosity," and the priority of "service" to his country,[52] as opposed to the Platonic vice and Christian sin of "avarice." Accordingly, the references to rational time in Jefferson's memorandum books hardly furnished the basis for modern economics as they did for the natural sciences. Instead, they served two social functions. First, Jefferson needed to record the exact dates of borrowing or lending money in order to organize his relations with creditors and debtors (rather than to have a financial overview of the total of his debts or the total of his outstanding money). This personal attention to time was of great importance in Virginian society and economics, which depended on an intricate web of debts and credits, as many planters were both creditors and debtors.[53] Second, the habitual exactness of Jefferson's financial accounts was linked to the larger aim of self-control. This problem surfaced conspicuously in the field of his personal expenses, where he may have found it most difficult to discipline himself.

The second social function of rational time, a cultural ideal of constant self-discipline, is condensed in the term *punctuality,* in its wider reference not only to temporal precision but also to a strict observance of rules, a penchant for formality, or a correctness of conduct in general.[54] The ideal of punctuality in Jefferson's culture can be interpreted in the terms of Max Weber's "innerweltliche Askese," David Riesman's "inner-directed character," and Norbert Elias's "Zwang zum Selbstzwang," specifically his "Zwang zur Langsicht."[55] Whether seen in the broader context of an increasingly secularized Protestantism underlying modern industrial capitalism, of a comprehensive internalization of values in a transitional society, or of affect control in the process of civilization, Jefferson's concern with a rational control of time was intertwined with his ideas about the control of the self in its relations to others. His preoccupation with what he called "a methodical distribution of our hours"[56] was a recurrent topic in his letters to young family members, whose habits, in terms of Riesman's "inner-direction,"[57] he sought to regulate while they were young in order to ensure their future happiness. The "methodical" use of time implied in these regulations displayed several characteristics of what Weber would refer to as "a consistent method of the entire life-style."[58] Yet Jefferson's methodical use of time is unlikely to have been identical with that of the famous contemporary who became, rightly or wrongly, Weber's chief witness. A brief comparison with Benjamin Franklin's statements on time might therefore clarify the contours of Jefferson's temporal discipline: to what extent could his rational time become money?

Rational Lives

Jefferson gave his most detailed explanations of what he called a "rational life" in letters to young family members.[59] Beginning in the 1780s, he admonished his daughters Martha and Maria, his nephew Peter Carr, his future son-in-law Thomas Mann Randolph, and later his numerous grandchildren to avoid the evils of idleness and ennui–"the most dangerous poison of life"–by a careful "distribution" of their time.[60] For this purpose he devised different types of day schedules, either in the form of a list reminiscent of the famous schedule in Benjamin Franklin's *Autobiography,* in the form of a questionnaire, or in the form of an argument, discussing different possibilities for organizing the day.[61] Jefferson assigned fixed hours not only for study but also for activities such as physical exercise. The modern division between labor and leisure, "work" and "free" time, that structured Franklin's approach, however, was far from self-evident to Jefferson. In Franklin's schedule, reading and music were leisure-time activities, strictly separated from an artisan's "work." For Jefferson, these activities could not as easily be distinguished from "work," because they were part of a comprehensive educational program for members of the Virginia gentry. Even if his young grandsons, for instance, also learned a "profession," such as law or medicine (as opposed to farming, which Jefferson understood as "unprofessional" by definition),[62] he regarded them principally as future country gentlemen and responsible citizens, whose manners and erudition could not be neatly separated from their "work."

On the whole, Jefferson's lifestyle exemplified a premodern unity of labor and leisure time. In a period of emerging modern notions of professionalism, however, there were occasional uncertainties in his terminology. One example is his changing attitude toward the place of letter writing in his routine. Shortly after his presidency, he described his correspondence ironically as his "trade" in Washington, while he expected it to be a mere "hors d'oeuvre" in the "country economy" at Monticello, where it could assume "no part of the regular routine of the day."[63] During his retirement, however, Jefferson's correspondence consumed more and more of his time, once more becoming the "regular routine" of his days. He accordingly began to regard it as "incessant labor" to be "drudging at the writing-table."[64] Jefferson tried to rationalize this "labor" by using copying devices such as the polygraph, which he called "the finest invention of the present age."[65] Yet for the Ciceronean farmer-statesman-philosopher, it was still difficult to draw the line between labor and leisure and to determine whether his correspondence belonged to the labor of a statesman-philosopher or to the leisure of a farmer. In early nineteenth-century Virginia the boundaries between "work" time and "free" time were still fluid, whether for the slaves, who by definition did not have any "free" time, or for their owners, who tended to regard their "unprofessional" farming as inseparable from their social status, hence ideally as continuous and interminable.

In this sense, Thomas Jefferson and his family were indeed "*always* busy . . . *always* doing something," as the Monticello overseer, Edmund Bacon, observed.[66] Yet their ideal of more or less constant activity could not be subsumed under a modern concept of labor. In his family letters, Jefferson recommended "industry" and "employment" of body and mind less for economic and financial reasons than for social ones:

> It is your future happiness which interests me, and nothing can contribute more to it (moral rectitude always excepted) than the contracting a habit of industry and activity. Of all the cankers of human happiness, none corrodes it with so silent, yet so baneful a tooth, as indolence. Body and mind both unemployed, our being becomes a burthen, and every object about us loathsome, even the dearest. Idleness begets ennui, ennui the hypochondria, and that a diseased body. No laborious person was ever yet hysterical. Exercise and application produce order in our affairs, health of body, cheerfulness of mind, and these make us precious to our friends.[67]

In this letter to his daughter Martha, Jefferson presented health, happiness, and the "esteem" of others as the most important aims of habitual industry. Economic or financial benefits from "laborious" behavior–"utility," in Jefferson's unspecific paraphrase–were at best supplementary points in his argument. This was also true of his letters to young men, for whom financial success might be expected to have been a more vital motive. Regarding both sexes, Jefferson's admonishments on a conscientious use of time were directed toward more comprehensive goals than mere financial profit.[68]

Since Jefferson warned against the Christian sin of sloth mainly because of its consequences for one's health and social relations, his idea of a "rational life" can be understood as a secularization of the Benedictine motto "Idleness is an enemy to the soul." This secularization, however, did not go as far as Franklin's blunt "Time is money," which transferred the Christian feeling of responsibility for the use of God-given time into what was becoming the amoral context of purely financial profit. While Franklin also cited health, honor, and happiness as incentives for proper time management, he often referred to the concrete financial gains deriving from "saving time."[69] Jefferson did not make this connection, and his argument for a methodical distribution of rational time was couched primarily in the terms of traditional morality and social responsibility. Unlike the *Poor Richard* almanacs, his self-consciously "disinterested" voice did not suffer a descent into the unpleasant details of pecuniary problems caused by "losing" or "wasting" time. His argument for propagating a "rational life" focused less on the appropriation than on the preservation of a social status that Virginia landholders had, in principle, already attained. It was socially less inclusive than Franklin's: Jefferson's recommendations of "industry" and "employment" to his family have to be read in the tradition of the *diligens paterfamilias,* as part of an identity already grounded in a socially privileged position.

Jefferson thus would not have been able to embrace Franklin's explicit equation of time with money, which would become basic to an industrial capitalism based on timed and specialized labor.[70] Stylistically, however, there were several points on which Jefferson's and Franklin's statements converged. Like many of their contemporaries, both authors liked to depict time as something extremely scarce and "precious."[71] Since it could be lost, wasted, or saved, time appeared connected to some idea of individual ownership. In Franklin's as well as Jefferson's writings, it figured as a limited and valuable commodity, to a greater or lesser degree at the disposal of rational human beings. Hence, despite his aristocratic aversion to "sordid avarice," there was a stylistic rapprochement between time and money in Jefferson's writings. Regarding time, Jefferson could indeed be, as his grandson put it, a "miser." In his educational letters, the urgency with which he insisted how "little time" was "left" to people thirty or even sixty years his junior is striking.[72] Since he perceived time as extremely brief and immensely valuable, great care had to be taken in using it economically. For if it was "lost," it was "never found again" and "lost for ever," in both Franklin's and Jefferson's words.[73]

There is something peculiar and, as remarked by Max Weber, quite "irrational" about the notion of ownership and "loss" of time in Western culture, especially when it is largely independent from a financial context (in which the loss of time could indeed mean a loss of money), as in Jefferson's case. An individual loss of money can easily be imagined, but how exactly can one conceive of a "loss" of time as such? Understood literally, where does it disappear to? And how can its "loss" be avoided or compensated? Many eighteenth-century texts were characterized by a particular emphasis on the scarcity of time and a feeling of obligation to make up for "lost" time. This emphasis was often tied to a notion of having come late, as in Voltaire's "Nous sommes venus tard en tout. . . . Regagnons le temps perdu."[74] Usually, the idea of "lost" time in the present worked with an explicit reference to the future. Time was "lost," in this sense, when it had become past without contributing to a better time to come. In their didactic writings, both Jefferson and Franklin founded their advice of a "methodical distribution" of present time on the recurrent theme of the necessity of providing for the future. The (illusory) implication seems to have been that by guarding rather than losing time, and by saving rather than consuming it in the present, one could ensure that more time was actually left and could be spent in the future. And a present loss of time could be compensated only by an ever more prudent time management in the future. As Jefferson told his nephew Peter Carr, "The way to repair the loss is to improve the future time."[75] Jefferson's day schedules were similar to Franklin's in their legitimization of temporal discipline by the invocation of a better future, as if the total amount of one's time, like money, could actually increase when saved or wisely invested. As has been criticized as characteristic of a time-schedule culture,[76] both men considered the future, in this context, as the standard for behavior in the present. In contrast to the past, the future was not yet "lost," as it could still be planned and manipulated, to some degree, by a methodical self-control and "distribution" of time in the present.

The idea that time was scarce subjectively increased the "tempo" of life.[77] In Jefferson's experience, this quickened pace of life may account for his tendency to expect important changes to happen "at some period not very distant," though actually long before they occurred: his recall from France, his retirement, and, over the last four decades of his life, his death.[78] The anticipation of death extended beyond his personal life; he even taught his eleven-year-old daughter Martha "always" to be "prepared for the end of the world: or for a much more certain event which is death."[79] What seems rather a dogmatic insensitivity in a letter to a girl who had recently lost her mother is not likely to have stemmed from a traditional memento mori in Jefferson's case. He clearly secularized Christian teleology in the letter:

rather than seriously discussing late eighteenth-century millennial expectations or generally alluding to the vanity of worldly endeavors, Jefferson explicitly denied "those foolish predictions" of an approaching end of the world.[80] Instead, he focused on the individual problem of death: "This must happen to all: it puts an end to the world as to us, and the way to be ready for it is never to do a wrong act."[81] Jefferson thus secularized death by discussing it in the context, not of attaining the grace of God in the next life, but of education in this one. He recommended the thought of death as an educational measure because of its linear structure: its basic future-orientation was supposed to teach Martha always to "consider before hand."[82] Whether the problem she encountered was simply a spelling error or a more consequential "wrong act," it was important to learn to think about it *before* it actually took place.[83] The point, in other words, was not death itself but her general "preparedness"–for the end of her life as well as for any other event. Jefferson was trying to teach Martha habitual foresight, which he considered to be a crucial attribute of an American in her position.

His argument for a "rational life," then, was an argument in between two worlds and two ways of thinking.[84] On the one hand, it still had vaguely religious overtones and did not acknowledge financial benefits to be gained from time thrift. For Jefferson the maxim "Time is money" would have come too close to his detested "avarice." Yet on the other hand, Jefferson's argument implied that time was a valuable commodity that had to be managed well in order to maintain a secular social status. This status was defined by an ancient ideal of "unprofessional" landholding, rather than professional work. The worldly goods that motivated "the subordination of short-term affects to the principle of habitual foresight,"[85] accordingly, were not professional success and wealth but a social position distinguished by "honor" and "esteem," in Jefferson's terms. Transferring a blend of God's foreknowledge and the foresight of Aristotle's just master[86] into the eighteenth century, Jefferson linked habitual foresight to a sense of social responsibility. Especially in a slave society, time had to be a privilege of the masters; slaves' "owning" their time and making their own plans for the future questioned the very principle of slavery.[87] Yet slaves were not the only group to whom Jefferson denied his ideal of foresight. Because of the approximative character of rational time, he assumed that all human beings–Virginia slaves, European aristocrats, and even American citizens–could differ considerably in the way they perceived time.

Chapter 2

Paternal Punctuality

Thomas Jefferson's fictional namesake Sir Thomas, English absentee slave owner, member of Parliament, and *paterfamilias,* is the character in Jane Austen's *Mansfield Park* (1814) who is portrayed with the most "correctly punctual habits."[1] When the novel is read as a reflection on time, the inhabitants of Mansfield Park exemplify how different ways of perceiving and dealing with temporality could characterize social relations in their transatlantic "world of changes." For instance, at the other end of the social and philosophical spectrum–as far removed from Sir Thomas's rational punctuality as possible among the protagonists of the novel–the young unmarried visitor to Mansfield Park, Mary Crawford, declares that she "cannot be dictated to by a watch." Between these two poles, Sir Thomas's second son, Edmund Bertram, is "not yet so much in love as to measure distance, or reckon time, with feminine lawlessness."[2] As the words "dictated" and "lawlessness" indicate, precision of time measurement has the potential to regulate social relations in the novel. At the end of *Mansfield Park,* the unequal access to a "lawful" knowledge of time largely parallels the hierarchies of age and gender, social status and economic independence. Written at a time when the ownership of clocks and watches was a luxury that was increasingly "democratized,"[3] the novel presents Fanny Price's punctual attention to "the great clock at Mansfield Park" as a reflection of her social aspirations to a higher status in the Bertram family.[4]

The two faces of Thomas Jefferson's historical Great Clock at Monticello, as described in chapter 1, conveyed different information. The perception of weekdays, minutes, and seconds was the privilege of those viewing the clock from inside the house, while those outside the house had to be satisfied with the sound of the bell or with the hour hand on the outer face of the clock. As the fictional and historical examples of the great clocks at Mansfield Park and Monticello illustrate, the emphasis on space in enlightened time epistemology had a social significance: the exactness of time measurement was related to who was allowed inside and who was left outside the great house, both literally and figuratively.[5]

Materializing the connection between time perception and (social) space, clocks and watches were fraught with symbolism. In Jefferson's universe, this symbolism was most portentous in the context of slavery. Thus, he had a tradition of giving his daughters and granddaughters a gold watch together with a slave as a birthday present when he considered them old enough to take on domestic responsibilities and train their own slaves.[6] As in the relationship between Robinson Crusoe and Friday ("I call'd him so for the Memory of the Time; I likewise taught him to say *Master*"),[7] the Jefferson women had to be able to calculate clock time if they wanted to master their own lives and become the mistresses of other human beings. When the Monticello slave James Hemings had the opportunity to become legally free in Paris, Jefferson bought him a watch,[8] certainly due to a greater practical need for temporal coordination in an urban environment but possibly also in recognition of Hemings's newly gained personal independence. In this sense, the possession of a watch symbolized the fact that Hemings received wages, making him the (legal) proprietor of his time. Slaves, by contrast, could not have "free" time; they were by definition excluded from "owning" their time to the point of frequently being ignorant of their age.[9]

Since timekeepers had the potential to signify someone else's ownership of their lifetime, it is not surprising that slaves might feel harassed by their masters' clocks. For example, Olaudah Equiano relates that he was originally afraid of his master's watch, fearing that its control over his life could go so far as to tell "the gentleman anything I might do amiss."[10] The quasi-human agency attributed to watches and clocks extended to contexts outside of slavery. Thus, Jefferson's silver pocket watch is supposed to have figured quite actively in the "midnight" appointments in 1801, allegedly used by Jefferson's new attorney general to interrupt Chief Justice John Marshall as he was writing out the last Federalist commissions.[11] The fear of the authority symbolically exerted by mechanical timepieces could also turn into aggression, at least among the (white) sons and grandsons of the Virginia revolutionaries, who reportedly used Jefferson's second "great clock," at the University of Virginia, as a favorite target for sharpshooting.[12] The Romantics, most famously the University of Virginia student Edgar Allan Poe, would tend to imagine clocks as dangerous machines identified with everything they despised about the previous generation. Jefferson may have encountered an early Romantic clock monster himself, in the novel *Wieland, or The Transformation* (1798), probably sent to him by its author, Charles Brockden Brown.[13]

As the contentious nature of these symbolic clocks may illustrate, critics as well as proponents of rational time tended to link it to questions of social control. For Jefferson and like-minded contemporaries, this control became all the more precious as they saw it endangered from different directions: socially, from above and from below; historically, by older and by younger generations. Punctuality and future-orientation offered them ways to combine the necessity of change with the hope for stability. Structured by rational time, their "world of changes" could be described as ordered and teleological, at least in the longer term. Conversely, the absence of a rational time perception could be used to define the limits of their ideal but endangered middle position. From the perspective of Euro-American writers in the eighteenth century, various people–whether aristocrats, servants, "savages," Orientals, black slaves, women, or children–could appear to live in another temporality,[14] a temporality in which they could be depicted as incapable of foresight, obedience to day schedules, or exact time measurement. Many texts of the period remained ambiguous concerning whether these exclusions were meant as affirmations or rather as expressions of a certain uneasiness and ambivalence in regard to rational time. In this sense, the projection of a divergent "temporal orientation" on specific groups was part of a critical discussion in the Enlightenment of a temporality that would be identified as typically "enlightened" in retrospect, although it would in fact only define a mainstream way of life well after the eighteenth century had ended.

Jefferson's concern about the proper attention to rational time centered on a relatively select group of people, all members of the Virginia planter elite. His first object of attention was his own behavior, to conclude from his near-obsessive record keeping and characterizations by contemporaries that echo the strengths and weaknesses of the fictional Sir Thomas's "correctly punctual habits." Jefferson's advice to young family members was modeled on the habitual punctuality of the *diligens paterfamilias,* which he himself aspired to. Since this persona did not have to be concerned with the acquisition of money to gain social status, it could not have served, in the same way as Franklin's writings, as an incentive to rational time management for urban merchants, artisans, or workers. Although Jefferson did not explicitly address these parts of society in his writings, he usually did not go so far as to use them as negative counterexamples to rational punctuality and foresight. This function was fulfilled mainly by two other groups: aristocratic ladies in Paris and black slaves in Virginia. As Jefferson aged, he began to criticize a third group, the "post-revolutionary youth" in America.[15]

French Ladies

In a letter to Anne Willing Bingham written at about the midpoint of his stay in Paris, Jefferson tried to engage his American correspondent in an argument about whether the American way of life was preferable to the European.[16] His own opinion in the matter can easily be surmised and is not difficult to discern in his introductory phrases: "You are then engaged to tell me, truly and honestly, whether you do not find the tranquil pleasures of America, preferable to the empty bustle of Paris. For to what does that bustle tend?" The exaggerated opposition between "tranquil pleasures" and "empty bustle," contrasting a way of life that was slow and happy with a life that was senseless and rapid, left little doubt as to where Jefferson's preferences were to be expected (which Anne Bingham, to be sure, would consider "rather overcharged" in her reply).[17] In the rhetorical question "For to what does that bustle tend?" Jefferson implied the thesis of his letter: the "bustle of Paris" was "empty" because it had no direction, no "future good" connected to it. To illustrate this point, Jefferson went on to describe what he considered the typical day schedule of a lady in Parisian society:

> At eleven o'clock, it is day, *chez madame.* The curtains are drawn. Propped on bolsters and pillows, and her head scratched into a little order, the bulletins of the sick are read, and the billets of the well. . . . If the morning is not very thronged, she is able to get out and hobble round the cage of the Palais royal, but she must hobble quickly, for the *coeffeur's* turn is come; and a tremendous turn it is! Happy, if he does not make her arrive when dinner is half over! The torpitude of digestion a little passed, she flutters half an hour through the streets, by way of paying visits, and then to the spectacles. These finished, another half hour is devoted to dodging in and out of the doors of her very sincere friends, and away to supper. After supper, cards; and after cards, bed; to rise at noon the next day, and to tread, like a mill horse, the same trodden circle over again.[18]

Although this description might sound like a criticism of one of the less respectable women in a Boucher or a Fragonard genre painting, the lady presented here can be presumed to have belonged to the French aristocracy or *haute bourgeoisie,* as Jefferson probably would not have admitted to an acquaintance with the Paris demimonde in a letter addressing Mrs. Bingham. What he discussed so contemptuously here was the way of life of the elegant and politically influential social circles to which he was trying to gain acceptance at this time, as American envoy and ambassador in Paris. Nevertheless, the passage can be read as describing the antithesis of the ideal day schedules he designed for his daughters. Jefferson's depiction of the French lady's typical schedule contains an analogous "distribution" of small temporal units (even to half hours) but contrary features in almost all other aspects. Rising at noon instead of sunrise, the lady is characterized by a general absence of order and cleanliness ("her head scratched into a little order"). She seems to lack health and strength: the "bolsters and pillows" combined with the habitual "bulletins of the sick" in the morning emphasize that disease was a constant presence also in her privileged part of society. The passive constructions in the initial phrases ("the curtains are drawn," etc.) intensify the impression of the lady's indolence and physical weakness. Most of all, she is ridiculed when he claims that she does not walk but either "treads" aimlessly like a horse or merely "flutters," "dodges," and "hobbles" around feebly on the high heels of the *ancien régime.*

The abundance of weak synonyms for a verb signifying a directed movement conveys the impression that this type of day schedule precluded any real "progress" on the part of either the lady herself or her society. Jefferson liked to use the verb *to hobble* to portray the weakness of an unprogressive Europe, as in the following remark made three decades later: "Old Europe will have to lean on our shoulders, and to hobble along by our side, under the monkish trammels of priests and kings, as she can."[19] Whether the image of hobbling was connected to the infirmity of old age or to the overrefined dress of a rococo aristocrat, it carried wider philosophical implications of decay and decadence. Like her entire continent, Jefferson's European lady was not progressing in a self-determined direction, but could at best "hobble along" or "hobble round the cage." Even though her life was characterized, like Jefferson's own, by the modern sentiment of an acceleration of activities–an impression reinforced by the staccato of hectic half sentences in the letter ("After supper, cards; and after cards, bed; to rise at noon . . .")–this "bustle" did not accelerate in order to lead more quickly into a better future. Its acceleration is portrayed as aimless and cyclical, always following, "like a mill horse, the same trodden circle over again." As Jefferson went on to explain his caricature, he introduced a surprising first-person plural into the letter: "Thus the days of life are consumed, one by one, without an object beyond the present moment; ever flying from the ennui of that, yet carrying it with us; eternally in pursuit of happiness, which keeps eternally before us. If death or bankruptcy happen to trip us out of the circle, it is matter for the buz of the evening, and is completely forgotten by the next morning."[20] The sudden seriousness of this general reflection on the passing of time, quite abruptly terminating the gossip about the frailties of French ladies, makes Jefferson's famous "pursuit of happiness" appear like rather a melancholic enterprise. Uncharacteristically perhaps, the pursuit of happiness was doomed always to remain a mere pursuit here for a group in which Jefferson at least rhetorically or temporarily felt himself included. While in Paris, he was part of a society that in his opinion merely "consumed" time (and everything else), without the perspective of future improvement. Relations in this superficial society were "thinned to cob-web by being spread over a thousand objects." And they were short-lived: apart from the spatially "thinned" social ties in the extended social web of an urban environment, individual affections were also extremely diminished temporally. The death or ruin of "very sincere friends" was forgotten as soon as it had lost its entertainment value ("the buz of the evening"). Yet although the memory of individual pain disappeared instantly, real happiness could not be achieved. Jefferson criticized an ephemeral lifestyle divested of both past recollections and future goals ("without an object beyond the present moment") as an "eternal" circling around a hopeless "pursuit of happiness," which was to be disrupted only by one's death or by the social death of bankruptcy. The paradoxical eternity of this pursuit was reflected in the circular structure of the letter, which Jefferson decided to end "where I began, like a Paris day."[21]

As he saw it, however, this unprogressive temporality was relative to one's position in space. Jefferson contrasted the "vanity of vanities" character of urban life in Europe with his version of a secular heaven: "In America, on the other hand, the society of your husband, the fond cares for the children, the arrangements of the house, the improvements of the grounds, fill every moment with a healthy and an useful activity. Every exertion is encouraging, because to present amusement, it joins the promise of some future good."[22] Comfortably back at home and back to his original topic of comparative women's habits, Jefferson reasserted the ideal of a "rational life" in which he tried to instruct his daughters at the time he wrote the letter–a bustling life which supposedly was not "empty," but "filled" with domestic employment and the "tranquil pleasures" of a rural environment (where the "grounds," like female selves, could be improved). Always oriented toward "the promise of some future good," this life contrasted sharply with the conspicuous consumption of time among the urban "artificial aristocracy."

It is striking that Jefferson used a female rather than a male aristocrat for his antithesis of the "rational life." In part, of course, the unflattering description of the French lady has to be explained in the context of the letter, as a *galanterie* to Mrs. Bingham, who was much favored by the comparison. Yet Jefferson's choice of a female character had a wider significance, as women were frequently excluded from a rational mode of time perception in contemporary literature.[23] During their pregnancies, to take the obvious example ironized in *Tristram Shandy,* they appeared subjected, not to the precision of clock time, but to the less predictable times defined by their bodies or their "nature." In Jefferson's description of the French lady's unsteady lifestyle, this biophysical temporality is reflected in his uncommonly specific references to her disorderly physical appearance ("propped on bolsters and pillows, and her head scratched into a little order"), the "tremendous" amount of time consumed by the care of her body ("the *coeffeur's* turn is come; and a tremendous turn it is! Happy, if he does not make her arrive when dinner is half over!"), and even the time taken up by her bodily functions ("the torpitude of digestion a little passed"). The details of this information would not have been strictly necessary to describe her daily activities, but the images evoked by her sluggish morning in bed, her lengthy coiffure, and her slow digestion reveal a lifestyle in which the intervals of time were structured by biophysical procedures. From Jefferson's perspective, the problematic aspect of the lady's time management was that she was not habituated to mastering her own "nature," in the sense that she was unable (or unwilling) to subject the physical needs of the moment to the abstract demands of "some future good."

An unprogressive temporality was thus condensed in Jefferson's satirical description of a woman's day schedule. However, this did not mean that he considered women to be innately incapable of organizing their lives according to rational time. When they had acquired the habits of "natural aristocrats" in the American countryside, they could lead "rational lives" as well as men. These lives surely would take place in a different sphere and would contain different activities,[24] but they could likewise be organized by rational punctuality and future-orientation. Within the limits of his conception of proportional equality, Jefferson appears to have had an environmentalist understanding of gender characteristics.[25] This held true also in the opposite direction. As the letter to Mrs. Bingham implies, a "Paris day" contained similar dangers for the men in European society, quite possibly even for the writer himself. Describing the melancholy of a hopeless "pursuit of happiness," after all, Jefferson used a general "we."

If he tended to speak of women when criticizing the unprogressive way of life in the Old World,[26] it was evidently less because he wanted to make a point of the inborn inferiority of their rational faculties than because the reference to women made it rhetorically easier to criticize artificial aristocrats in European courts and cities. Ridiculing European noblewomen, Jefferson could build on a number of discursive links between the *noblesse d'épée* and femininity in Enlightenment culture. The two groups could be criticized for similar deficiencies, many of which were connected to the perception of time. From a republican perspective, the "feminine" lifestyle of the nobility was characterized by such time-related aspects as their capricious sensuality rather than constant conjugal love, their cultivation of passivity rather than civilized activity, their short-lived embellishments rather than long-term improvements, their consumption in the present rather than production for the future, their cyclical decadence rather than linear progress, and so on. Blue-blooded women could thus be seen as representatives of an outdated way of life that also included men—even outwardly "effeminate" men, in courtier fashion adorned with laces, wigs, and makeup (an appearance that Jefferson's own resembles most in the Mather Brown portrait painted in London in 1786, before he began to be represented in more robust republican fashion with his hair *au naturel,* as in the 1788 painting by John Trumbull). The stock character of the immoral noblewoman, scheming to destroy many a hero or heroine in eighteenth-century dramas and novels, could be employed to personify the problems of the entire *ancien régime:* the court intrigues, the waste of time and money by the aristocracy and clergy, and eventually also the downfall of the monarchy.

Politically, the representative function of the immoral noblewoman became most useful to Jefferson in his characterization of the French queen. It allowed him to join many of his contemporaries in the claim that it was Marie Antoinette who was personally responsible for the radical turn of the French Revolution and the ensuing chaos, the temptations of her rococo garden ultimately causing the monarchy's accelerated fall from grace:

> This angel, as gaudily painted in the rhapsodies of the Rhetor Burke, with some smartness of fancy, but no sound sense was proud, disdainful of restraint, indignant at all obstacles to her will, eager in the pursuit of pleasure, and firm enough to hold to her desires, or perish in their wreck. Her inordinate gambling and dissipations, with those of the Count d'Artois and others of her clique, had been a sensible item in the exhaustion of the treasury, which called into action the reforming hand of the nation; and her opposition to it her inflexible perverseness, and dauntless spirit, led herself to the Guillotine, & drew the king on with her, and plunged the world into crimes & calamities which will forever stain the pages of modern history. I have ever believed that had there been no queen, there would have been no revolution.[27]

In opposition to the praise for Marie Antoinette in Edmund Burke's *Reflections on the Revolution in France,*[28] Jefferson could see in the queen at best a fallen angel–he liked to reserve truly angelic qualities for American women[29]–or rather as an Eve- or Helen-like figure who had fatally "plunged the world" into the misery of war and terror. Coming from the shepherd paradise of her artificial hamlet and her eternal "pursuit of pleasure," she had also plunged the world into the radical changes of a revolutionary temporality, "into the crimes & calamities which will forever stain the pages of modern history."

The king's behavior, by contrast, affected neither the world nor the times; he merely "plunges deeper and deeper" himself, as Jefferson wrote to John Adams in 1787.[30] Three decades later, in Jefferson's *Autobiography,* Louis XVI still appears as a "passive machine," anachronistically drawn to the guillotine in the queen's wake, a victim of her "inflexible perverseness, and dauntless spirit." Husband to "a Queen of absolute sway over his weak mind, and timid virtue," the king was presented by Jefferson as a monarch whose physical and political impotence made him appear almost as powerless as one of his own subjects.[31] The inversion of modern gender roles in the royal couple had a deep historical significance for Jefferson even before he knew the outcome of the French Revolution. Writing to George Washington in 1788, he interpreted the "omnipotence" of the "influence of women in government" as the true cause for "the desperate state" to which he saw things generally "reduced" in France.[32] Jefferson saw an entire country in decline when its representative women remained, as did Marie Antoinette, "firm" and "inflexible" in their resistance to "the reforming hand of the nation." From his perspective, the queen's inflexibility had its root in her traditional aristocratic pride and "dauntlessness." Her lack of rationality surfaced not only in her courage but also in an aggressive sensuality that was entirely unmindful of future consequences–in her "eager" pursuit of pleasure, her contempt of "restraint," and her fatalistic gratification of her "desires." When Jefferson thus characterized the queen in his *Autobiography,* the sexual implications of the "stain" he found she had caused "in the pages of modern history" had long since reached their historical climax in the allegations of incest with the dauphin in her trial in 1793. The fact that he continued to employ this language shows how disastrous the present-centered femininity of the *ancien régime* appeared to him. Marie Antoinette's temporal orientation toward an outdated past of aristocratic privilege and an immoral present of sensual pleasure continued to haunt Jefferson, fundamentally threatening to subvert the progress he envisioned for the future.

American Slaves

The second group that Jefferson identified with a present-centered way of life was located at the opposite end of the social spectrum in the transatlantic world of the 1780s. In a passage under query 14 of the *Notes on the State of Virginia* that has become notorious for its expressions of racial prejudice, Jefferson ascribed to black slaves in Virginia several characteristics already familiar from his criticism of European aristocrats.[33] Although analogies between privileged Europeans enjoying a life of luxury and American slaves enduring a life of drudgery may at first seem inappropriate, Jefferson's depictions of the two groups share important aspects, sometimes even expressed by the same terms and images. As the image of the "mill horse" in the letter to Anne Bingham demonstrates, it was plausible for Jefferson to illustrate the habitual indolence of decadent European aristocrats by a simile conveying a never-ending cycle of hard physical work. One possible explanation for this paradoxical comparison is that the two extremes, on the one hand an identity based on the complete absence of work and abundance of property and on the other hand an identity based on the constant presence of work and the complete lack of property, implied a similar temporality in Jefferson's thinking: like the continual work cycles of a mill horse, the routines of both groups were characterized by the impossibility of self-determined progress. In this sense, they were equally monotonous. Jefferson generally identified prerevolutionary society with "monotony," as when he referred to "the quiet and monotonous course of colonial life" and the "dull monotony of colonial subservience."[34] He regarded both established aristocracy and chattel slavery as outdated institutions belonging to the Old World, where property laws were "founded in force, and not in conscience."[35] Since Jefferson associated both groups with political monotony, Marie Antoinette's sensuality, bravery, and lack of "restraint," as well as the "stain" caused by her behavior, had distant echoes in his remarks on Virginia slaves.

Under query 14, Jefferson began his description of what he perceived as the "physical distinctions" between whites and blacks by diagnosing a deep-seated "monotony" specifically in the physical characteristics of blacks. In his opinion, "that eternal monotony, which reigns in the countenances, that immoveable veil of black which covers all the emotions" was aesthetically inferior to the appearance of whites, which he contrastingly defined by movement and change: apart from their "flowing hair," whites were distinguished by the ability to blush and to express "every passion by greater or less suffusions of colour." Thus whites, unlike blacks, were innately exempt from monotony in their outer appearance, displaying a greater potential for differentiation and change. For Jefferson, the trouble with the color black was that it was "fixed" in nature.[36] This perceived physical monotony had a significant cultural counterpart: supposedly, blacks had not yet risen on the civilizational scale of the arts and sciences. Despite the benefit of exposure to Euro-American civilization, as Jefferson chose to see it, black American slaves had not progressed even to an "elementary" stage in painting or sculpture, or "above the level of plain narration" in oratory.[37]

Jefferson insisted, in particular, on the deficiencies of the literature written by contemporary British and American blacks who lived outside plantation slavery, locating their works at a low point in the development of the art–"below the dignity of criticism" and "at the bottom of the column." The best he found to remark about the literary qualities of Ignatius Sancho's *Letters* was its "easy and familiar" style. Combined with the author's "strong religious zeal," this lack of sophistication seemed to place his work, like Phyllis Wheatley's religious poetry, at best at an earlier, more primitive stage in literary history.[38] What prevented Sancho from rising to the level of "the writers of the race among whom he lived," in Jefferson's opinion, was his lack of self-mastery: "But his imagination is wild and extravagant, escapes incessantly from every restraint of reason and taste, and, in the course of its vagaries, leaves a tract of thought as incoherent and eccentric, as is the course of a meteor through the sky. His subjects should often have led him to a process of sober reasoning: yet we find him always substituting sentiment for demonstration." This literary criticism was based on Jefferson's "conjecture" and "suspicion only" that blacks were by nature equal to whites in their moral faculties but inferior in their rational and imaginative ones. Sancho's letters could "breathe the purest effusions of friendship," but when it came to "sober reasoning," the writer had to substitute "sentiment" for (rational) "demonstration." To conclude from the implications of words like "extravagant," "vagaries," and "eccentric," this sentiment was momentary and accidental rather than stable and predictable. Jefferson used an image from astronomy that would also become important for his historical thought–the erratic course of a meteor–to illustrate the irrationality of Sancho's thought. The contrast between meteoric force and civilizing "restraint" was reinforced by another metaphor: Jefferson's words likened Sancho's untamed imagination to a wild beast opposing domestication, monotonously trying again and again ("always" and "incessantly") to "escape" from rational forms of control.[39]

Already familiar in connection with Marie Antoinette's extravagances, Jefferson's thoughts on literary "restraint" had an explicit physical context. Not incidentally, his discussion of Sancho's and Wheatley's writings was framed by remarks on black sexuality. A few lines after his criticism of Sancho's *Letters,* the term appears again in the following context: "But in this country the slaves multiply as fast as the free inhabitants. Their situation and manners place the commerce between the two sexes almost without restraint."[40] Apart from the use of this observation for a favorable comparison to ancient Roman slavery and for his argument that the fast growth of the black American population meant that the slaves' condition was "mollifying,"[41] the alleged lack of sexual restraint among American blacks had direct consequences for Jefferson's evaluation of their literary talents. As he explained: "Love is the peculiar oestrum of the poet. Their love is ardent, but it kindles the senses only, not the imagination."[42] It followed, for Jefferson, that there were as yet no real poets among blacks. Both sexual and literary restraint required a superior future-orientation enabling lovers and writers to postpone the gratification of their desires. Restraining one's sexuality for the sake of civilized love thus functioned on the same principle as restraining one's literary imagination in order to write enlightened poetry, and in Jefferson's eyes blacks were largely unable to do either.

The temporal essence of Jefferson's neoclassical restraint, in other words, was the subordination of the inclinations of the moment to the demands of "some future good." To Jefferson, blacks appeared as creatures of the present without talent for restraint or forethought. More immediately than whites, they were linked to a tangible nature, like children partaking more of Lockean "sensation than reflection."[43] In this characteristic, they could be seen to approach the construction of Rousseau's "natural man," whose foresight was so limited that he was unable to think further than to the end of the day. Unlike Rousseau's construct, however, the lack of foresight Jefferson "observed" among blacks was an attitude connected to social activities. A slave, according to Jefferson, could be "induced by the slightest amusement to sit up till midnight, or later, though knowing he must be out with the first dawn of the morning." This criticism, already become a commonplace in descriptions of American slave life in Jefferson's period,[44] was paralleled at the other end of the social spectrum by his reference to the French aristocrats' nocturnal "gambling and dissipations." Since both unenlightened groups preferred nighttime activities, they were inclined to sleep during the day. In addition to criticizing the unnamed French lady for sleeping late in the letter to Mrs. Bingham ("at eleven o'clock, it is day, *chez madame*"), Jefferson listed "their disposition to sleep when abstracted from their diversions" as a weakness prevalent among Virginia slaves.[45] The insinuation that blacks could sleep peacefully wherever they fell to the ground would become a racist cliché in the nineteenth century,[46] but in the eighteenth century it still had a wider meaning. In his *Essay Concerning Human Understanding* Locke had referred to sleep in order to exemplify the breakdown of the normal time perception.[47] In this more general context, Jefferson was able to use the topic of sleep to illustrate the temporality he associated with French aristocrats, with Virginia slaves, and eventually also with his own retirement years.

His emphasis on time perception enabled Jefferson to assume two further similarities between slaves and aristocrats. The superficiality of social ties in Paris, where personal catastrophes were supposedly "forgotten" on the next morning, was paralleled by the "transient" emotions Jefferson found it convenient to attribute to blacks.[48] Moreover, the traditional aristocratic attribute of courage, which he also granted Marie Antoinette, was one of the few in which he admitted that blacks might excel American whites. As in the case of the French queen, however, theirs was not a positively connoted type of courage. Jefferson understood it mostly as a kind of stupidity, an inability to calculate future risks. As he saw it, it was only their "want of forethought" and incapability of "seeing a danger till it be present" that made blacks "at least as brave, and more adventuresome" than whites.[49]

Contemporary African Americans themselves, if they were in a position actively to participate in this discourse, tended to make claims to the very forms of rational time management that Jefferson's "suspicion" denied them. The self-taught mathematician and astronomer Benjamin Banneker sent Jefferson a manuscript copy of an almanac that he had "calculated . . . with correctness and accuracy." Using his almanac to prove that he was endowed with the "same faculties," Banneker grounded his argument specifically in his talent for future-oriented study. Referring to the self-evident truths of the Declaration of Independence, "worthy to be recorded and remember'd in all Succeeding ages," Banneker achieved two rhetorical goals at once: he exposed the double standard of the American revolutionaries and the fundamental injustice of American slavery, while demonstrating that his mind was able to extend to the past as well as to the future, to the events of the Revolution and to the future significance of the Declaration. Through his almanac, his argument, and the construction of his argument, Banneker adeptly revealed as "prejudice" and "absurd and false ideas and oppinions" Jefferson's suspicions that blacks were creatures solely of the present.[50]

Jefferson's first short answer to Banneker's claim to a rational time perception was rather civil: he spoke of blacks as "our black brethren"–a formulation that he usually reserved for whites and Native Americans and that in relation to Banneker would later cause him accusations of "fraternization" by southern Federalists–and expressed wishes for "raising the condition both of their body and mind to what it ought to be." Whether mainly to promote Banneker's cause or to exhibit his own open-mindedness on the issue, he also sent the almanac to one of the foremost French opponents of slavery, the Marquis de Condorcet.[51] Annoyed by the later publication of his exchange with Banneker, however, Jefferson became more reserved over the years. Rather than looking forward to more examples among blacks proving "that the want of talents observed in them is merely the effect of their degraded condition, and not proceeding from any difference in the structure of the parts on which the intellect depends,"[52] Jefferson later reemphasized his "doubts" about racial equality expressed in *Notes on Virginia.* Never wavering in his belief that black slaves lacked the crucial attribute of foresight, whether he considered this deficiency to be environmentally conditioned (as in a letter to the abbé Grégoire) or an inborn characteristic (as in a letter to Joel Barlow) came to depend on the context and the correspondent. In both cases following the principle, as he explained to Barlow, "that nothing was or is farther from my intentions, than to enlist myself as the champion of a fixed opinion," he remained consistently ambiguous on the crucial question whether the supposed lack of foresight was the result of nature or nurture.[53] Writing to Edward Coles in 1814, he framed the problem in paternalistic terms: "For men probably of any color, but of this color we know, brought from their infancy without necessity for thought or forecast, are by their habits rendered as incapable as children of taking care of themselves."[54] For Jefferson, the slaves' childlike inferiority consisted specifically in their lack of foresight. In this respect, his argument was part of an ancient tradition: intellectual inferiority and a lack of foresight had also been defining characteristics of Aristotle's "natural slave," a person not merely enslaved by law but also "naturally" fit for slavery. Given his enlightened background, however, Jefferson's views had to be more complicated than those expressed in Aristotle's *Politics.*[55] Hence, he used the environmentalist argument of the slaves' upbringing to explain their perceived inferiority in the letter to Coles, all the while insinuating an updated version of the ancient tradition by establishing a contrast between probability (all men would "probably" be inferior under the same conditions) and certainty ("but of this color we know"). By alluding to both possibilities, Jefferson typically left open the question whether he considered the lack of foresight to be "fixed in nature" or capable of improvement after the removal of slavery. This ambiguity makes it difficult, if not impossible, to decide whether his remarks on black inferiority revealed an essentialist "race" or (merely) an environmentalist "class" prejudice.

Jefferson's tendency to evade commitment is reflected in his frequent allusions to sexuality, a domain where questions of nature and culture were so intimately linked that it was particularly difficult to keep them apart. In the passage in the *Notes* establishing the transition from blacks' physical to their mental characteristics, Jefferson wrote: "They are more ardent after their female: but love seems with them to be more an eager desire, than a tender delicate mixture of sentiment and sensation."[56] In this statement, it is not entirely clear whether he referred to innate or acquired characteristics. The assumption that black slaves could only enjoy spontaneous physical love but were unable to achieve constant sentimental love could be grounded in a more carnal nature, but it could also be caused by cultural factors, such as their particular "situation and manners" (whether originating in Africa or in American slavery). A present-centered sexuality, in any case, was not a characteristic Jefferson reserved for blacks: he lamented a lack of restraint, even juxtaposed in one sentence with suggestive words like "eager" and "desire," not only when describing enslaved black men but also when referring to the queen of France.

Animals

One of the statements under query 14 of *Notes on the State of Virginia* that has considerably disturbed modern readers is Jefferson's analogy between the supposed sexual preference of black men for white women and the fashionable idea of orang-outangs' mating with African women. Black men desired white women, as he put it, "as uniformly as is the preference of the Oran-ootan for the black women over those of his own species."[57] While this faith in the anthropomorphous qualities of orang-outangs may in part have been used by Jefferson as a vehicle to show off his knowledge of what was then a common (although not undebated) assumption in natural history, it is certainly one of his most problematic statements regarding blacks, especially judged by its consequences for the racist anthropology of the nineteenth century.[58] Considering, moreover, Jefferson's long-term relationship with his slave Sally Hemings,[59] his musings on the sexual desires of orang-outangs acquire rather an embarrassing aspect in retrospect, although to be sure, his warnings against miscegenation in *Notes on Virginia* were written before this relationship began. Locating blacks at a lower point in the "great chain of being" than whites and increasing their proximity to animals, Jefferson quite seriously referred to a possible crossing of the boundaries of "species," even leaving open the question of polygenesis. His train of thought thus started with his slaves' sexuality and ended with an awkward "philosophical" statement about their status in natural history, suggesting a lower "rank in the scale of beings which their Creator may perhaps have given them."[60]

Jefferson's motivation for such protoracist reflections needs to be analyzed, first of all, in the context of his views on slavery. To judge from his argument for gradual abolition in the *Notes* and elsewhere, reducing the distance between blacks and animals provided him with an indirect rhetorical advantage. Jefferson wished to argue for a temporary maintenance of slavery, while being unable to legitimate the institution as such, whether by the nature of the slaves or otherwise. Unlike later racists, he was convinced that he could not use the allegation of a proximity between blacks and animals to justify the principle of slavery.[61] Nevertheless, their supposedly lower status in the "great chain" at least implied the possibility that their perceived inferiority was grounded in natural rather than in civil history. To some degree, civil history seemed thereby relieved from the burden of responsibility for the "degraded condition" of the black slaves.[62] If their situation was understood as more innate than man-made, agents of civil history trying to improve their lot could only expect moderate success in the future. According to this train of thought, abolition was still necessary but could only be imagined as a long-term measure, leaving time for slaveholders to use their foresight to plan ahead for their slaves.

For an argument of gradual abolition, one did not have to insist that black slaves were "by nature" inferior to their masters. In a text that Jefferson began to translate in Paris, for example, the Marquis de Condorcet also offered a long-term emancipation program, although he–in contrast to Jefferson–was convinced that black and white "brethren" were by nature absolutely equal in their intellectual abilities.[63] If blacks were assumed, as in Condorcet's text, to be merely inferior by nurture, due to their habits in slavery, it was still plausible for contemporaries to argue that the slaves were temporarily in need of a more experienced master. To strengthen his rhetorical position, it was therefore sufficient for Jefferson to leave open the possibility that the "natural" characteristics of blacks might perhaps parallel their inferior social situation in slavery. This move had two benefits: On the one hand, Jefferson could display his enlightened open-mindedness toward the results of future "observations,"[64] a future-oriented attitude that need not have been insincere, as it was basic to his empirical frame of mind. On the other hand, his insinuation of the mere possibility of an innate intellectual inferiority would contribute to making his readers (and himself) more comfortable with the temporary continuation of slavery. For if the black slaves were depicted as "naturally" dependent, like children or domestic animals, on the superior foresight of their (benevolent) owners, the postponement of abolition would appear, if not more justified, at least less cruel.

This way of thinking would have serious consequences in the following generations, as it anticipated the essentialist racism of the nineteenth century. That Jefferson's own thought was not consistently in this direction may be suggested by the fact that he found analogies to domestic animals not only among black American slaves but also among the European elite. Explaining the decadence of the European dynasties at the time of Napoleon's victories, Jefferson wrote:

> Now, take any race of animals, confine them in idleness and inaction, whether in a stye, a stable, or a state room, pamper them with high diet, gratify all their sexual appetites, immerse them in sensualities, nourish their passions, let everything bend before them, and banish whatever might lead them to think, and in a few generations they become all body and no

> mind: and this, too, by a law of nature, by that very law by which we are in the constant practice of changing the characters and propensities of the animals we raise for our own purposes.[65]

Thus, the opposite of Jefferson's "rational life"–inactivity combined with immediate sensual gratification–would cause mental degeneration among all ranks of society. In combination with the cultural "practice of Kings marrying only into the families of Kings," this environmentalist explanation also took an essentialist turn, typically mixing, as in Jefferson's descriptions of American blacks, inborn and acquired characteristics. Hence, Jefferson could portray the queen of Portugal, for instance, as "an idiot by nature." In the "Natural History of Kings and Princes," whose publication Jefferson proposed in another letter, mental debility could pass from being environmentally conditioned to becoming innate.[66] As he expressed it, a monarch's "race" could change in time and "wear out." And worn-out races, it could be assumed, were naturally unfit for government.[67]

In his insistence that even the most blue-blooded "races" of Europe, like all races in the "animal kingdom," were subject to natural processes of degeneration, Jefferson pursued a rhetorical aim that was in one sense different from his suggestion of a proximity between blacks and animals and in another sense similar to it. His comparisons between monarchs and animals were supposed to diminish the veneration for European dynasties: they inverted the doctrine of divine right, portraying European princes as incapable of mastering even themselves, let alone God's creation in general or their subjects in particular. As Jefferson dressed such far-reaching claims in the terms of natural history, his style was clearly hyperbolic, contrasting sharply with his attempt at an impartial tone in his discussion of blacks as "subjects of natural history" under query 14.[68] Despite all his disgust for George III, for instance, Jefferson would probably not have objected to granting him at least a vegetable status in a sober context. Yet in a more playful mood, he could go even further down the scale of natural history by locating the Hanoverian kings at a lower "rank in the scale of beings" than all humans, animals and plants together, suggesting to "drop these odious beings and pass to those of an higher order, the plants of the field."[69]

Jefferson thus employed the language of natural history to characterize Old World monarchs as well as New World slaves, but this language mainly served satirical ends in the European context. Members of the nobility could therefore appear at even lower points in the "great chain of being" than people of African origin: if the latter were supposed to be occasional objects of desire for orang-outangs, they at least did not have to look up to "the plants of the field" as beings of a higher order. Especially after the achievement of American independence, Jefferson could permit himself more humorous exaggeration when referring to European monarchies than when discussing the unresolved issue of slavery, in which he was implicated, as he realized, on the wrong side.[70] His "suspicion only" of a closer proximity between animals and blacks was an important argumentative auxiliary in his long-term emancipation program: he needed it to render more plausible his argument that American slaves could not, for some time to come, make plans for themselves.

What was basic to both analogies, however, was a belief in the unity of the creation, an enlightened conviction that universal laws of nature were at work in stables and staterooms, influencing humans and animals alike.[71] Even the satirical comparisons to the nobility would not have made sense without the serious assumption that all human beings, regardless of color and social situation, were in danger of mental degeneration if they did not sufficiently restrain their natural inclinations. On this supposition Jefferson used natural history as a universally applicable tool to relieve civil history of some of its power to define social relations and political institutions. By attributing characteristics of a present-centered way of life to both the European nobility and American slaves, he excluded them, on protobiological grounds, from his vision of a progressive American future. From the vantage point of a "civilized" middle class, both groups were devalued in their animal-like qualities: American blacks compared to orang-outangs represented the perils of undercivilization, while European aristocrats compared to fat domestic animals stood for those of overcivilization. These dangers appeared all the more imminent as both groups were characterized by an unrestrained sexuality, threatening to impair, from below and from above, the personal integrity of republican citizens.

In Jefferson's writings, this regressive sexuality tended to be illustrated by feminine attributes in the case of European aristocrats and mostly by masculine attributes in the case of American slaves. While Jefferson regarded American men as in danger of being seduced by "the voluptuary dress and arts of the European women,"[72] it may be speculated that it was rather a rival *male* sexuality that men like Jefferson feared most in American blacks. Both forms of sexuality, in any case, threatened to cause some sort of national "stain." In European dynasties, women like Marie Antoinette could cause a "stain" in "the pages of modern history" if they produced illegitimate heirs. For an American slave, leaving a trace in the pages of history was more difficult. Yet "he" also threatened to leave his print on the future nation: while the black population, or "this blot in our country," as Jefferson saw it, was increasing "as fast, or faster, than the whites," an American slave could do so by "staining the blood of his [*sic*] master."[73]

Yet another anxiety underlay Jefferson's worry about a middle-class purity soiled by national "stains" and "blots." Physically powerful slaves and politically powerful aristocrats threatened to invert a republican order not only on a private but also on a public level. There were moments when Jefferson feared an "exchange of situation" between masters and slaves[74] and others when he was alarmed by the prospect of "monarchising" powers' using their feminine charms to "seduce" or "bewitch" upright republican citizens in America.[75] Especially in the former case, he considered a complete inversion of roles "among possible events" and even "probable" in the future.[76] While apocalyptic scenes of role inversion were a frequent topos in eighteenth-century polemics against slavery,[77] the fear of a devastating race war was a real one for Jefferson. It was basic to his argument for gradual abolition and colonization that he hoped to evade two kinds of civil wars: those that would be the outcome of too early an emancipation and those that would result from one coming too late.[78] Jefferson regarded the possibility of such wars as an existential threat to the survival of the republic, worrying that they would contribute to the "extirpation" of the white masters and even to "the evanishment of the human race."[79] This anxiety can in part be explained by his conviction that slave revolts were legitimized by a higher law.[80] Within an ordered universe, where everything–the slaves' legitimate desire for freedom included–had its purpose, he expected slave insurrections to take place with certainty at some point in the future. No doubt they would begin with small, local slave conspiracies and uprisings that would proceed, "in the order of things," like the method of induction from the particular to the general, and finally result in an all-encompassing race war. The fear that the American slaves would eventually be victorious was grounded not only in his insight that they were on the legitimate side of this contest–natural "enemies" whose "recollections" would always legitimize them to fight against "despots" who had mere "prejudices" on their side[81]–but also in his fantasy that their lack of future-orientation had resulted, as had that of Old World aristocrats, in an exceptional amount of primitive courage. To Jefferson's mind, once an open war between white masters and black slaves had begun, it could not be resolved by a peaceful compromise but would probably result in the death or enslavement of one side, this time probably that of the whites.

Although in this form unknown to Jefferson, it was a comparable kind of existential courage–the ability to overcome one's fear of death in the moment of primal combat–that originally distinguished the master from the bondsman in the writings of his contemporary Hegel.[82] Jefferson probably would not have appreciated Hegel's philosophy, as it exemplifies what he criticized as excessive abstraction in strict philosophical "systems."[83] Hegel's condensation of the revolutionary experience, however, drew on a shared fund of ancient and modern texts. Summarizing the paradoxes of numerous master-servant relationships in Enlightenment literature, his abstractions often revealed tendencies in discussions of servitude that were also relevant to Jefferson. His antithetical image of master and bondsman, considered on its most representational level, was therefore not too far from the actual experience of republican slaveholders like Jefferson,[84] who were at least partially aware that in their own fight for liberty they could "not imagine a single argument in their support which their own daily practice does not bely."[85] Concerned that the Virginia slaves would have both legitimacy and fortitude on their side in a possible contest, Jefferson perceived, in almost Hegelian terms, the first tendencies of a dialectical inversion of master and slave "spirit" in America when he claimed that the "spirit of the master is abating, that of the slave rising from the dust." He appears to have been terrified by the possibility of this "change," at least if it happened without the "consent of the masters" and thus eluded their control.[86]

Not incidentally, Jefferson's impassioned warning of a possible racial war in Virginia was interrupted by a remark criticizing the lack of "industry" among the masters, resulting in the apprehensive observation "that of the proprietors of slaves a very small proportion indeed are ever seen to labour."[87] In a Hegelian reading (rather than a classical republican interpretation)[88] of Jefferson's text, the fact that the masters were existentially dependent on their slaves' labor would have contributed substantially to the fear of an inversion of roles. While the master in Hegel's abstraction, roughly paraphrased, became by definition a passive consumer of the things of nature, the bondsman actively transformed the things, doing so to gratify not his own but the master's desire. Based on this self-restraint, it was only the bondsman who could look forward to a better future, dynamically moving the conflict to the higher level of a mutual recognition of the other's freedom. At least in Hegel's early writings, this synthesis was not achieved primarily, as Jefferson hoped in the case of American slavery, by a rational "consent of the masters." Rather, as in Jefferson's worst apprehensions, its realization at first had to go through a phase of violence and terror initiated by the former slaves.

While in Paris, Jefferson was shocked by reports that his fellow Virginians were returning to their former aristocratic habits of luxury and dissipation.[89] If Virginia citizens indulged in a present-centered lifestyle, unable to restrain their passions for the sake of "some future good," they would in the end resemble Hegel's master: after winning the original combat, passively consuming the produce of their slaves' labor and only waiting for them to kill or enslave them in return. As Jefferson knew very well himself, the slaveholders' lack of foresight on the prosaic level of economic and financial matters alone sufficed to "enslave" them again to their British creditors.[90] In his effort to consolidate the precarious middle position of Virginia republicans rhetorically, his insistence on the animal-like lack of restraint among American slaves and European aristocrats almost became a philosophical and psychological necessity. While admonishing his fellow Virginians to lead "rational lives," he sought to convince himself and others that all groups threatening to invert the republican order were, unlike Hegel's bondsman, only endowed with static and unprogressive characteristics: Virginia slaves were unmindful of the future, unable to restrain their physical desires, to contribute by their work to the American arts and sciences, and even to change their "monotonous" physical appearances. And the self-indulgent lifestyle of the European nobility, as Jefferson loved to point out later, had by Napoleon's time even made them physically incapable of fulfilling the master's traditional task of fighting.

Thus, Jefferson tended to identify as present-centered "animals" people who were either clinging to traditional privileges or beginning to claim universal rights of political participation, which he regarded as incompatible with republican principles or at least with the present state of the "public mind" in America. One way to depict, at the same time, American slaves and members of European dynasties as incapable of (self-)government was to imply that they were, due to their "irrational" time consciousness, equally unable to effect secular progress. Jefferson consequently linked American progress to an identity that tended to be republican, white and masculine rather than blue-blooded or enslaved, black or feminine. However, this division was by no means consistent. As Jefferson grew older, he became particularly disappointed with the next generation of American citizens: he criticized the "post-revolutionary youth" not only for their proto-Emersonian intellectual "self-sufficiency"[91] but also for a lack of temporal discipline exasperatingly similar to the habits of animal-like monarchs and slaves.

Fathers and Sons

In Jefferson's "world of changes," to close the circle to *Mansfield Park*, the achievement of enlightened foresight was anything but guaranteed. Like Jane Austen's fictional Sir Thomas, the historical Thomas Jefferson often saw his paternal punctuality in danger of becoming transformed by immediatist influences. In the year 1814, when the novel was published and Sir Thomas's fictional troubles with the next generation began, Jefferson was approached by a younger man asking for his paternal advice: his neighbor Edward Coles wanted to take immediate action against slavery in Virginia and asked Jefferson, "the revered father of all our political and social blessings," for support. In his response, paternally trying to persuade the younger man of his own more long-term gradualist position toward slavery, Jefferson made several references to time to affirm the authority of his opinions. Beginning with the claim that "time has only served to give them stronger root," he repeatedly mentioned his age and experience to assert a superior knowledge of past, present, and future. It was this temporal wisdom that he most explicitly denied to the people chiefly concerned by his gradual emancipation plans. The slaves, as he described them here in paternalistic manner, were deficient in "thought or forecast" and therefore as "incapable as children of taking care of themselves."

This assertion of paternal authority, however, was not limited to the slaves alone. In this letter Jefferson established a similar rhetorical opposition between a father- and a childlike figure solely among Virginia citizens. He built his argument also on the contrast between the temporal consciousness of the writer and that of the recipient of the letter, between the paternal foresight of Jefferson's generation and the disappointing present-centeredness of Coles's "younger generation." His own generation of revolutionaries, as he saw it, had managed to progress beyond the "monotonous course of colonial life" *their* fathers had been leading (figure 2). In contrast to this, the postrevolutionary generation, although "receiving their early impressions after the flame of liberty had been kindled in every breast," were failing to move human progress further, lost as they were in "an apathy unfavorable to every hope."

Jefferson disparaged two different approaches to slavery among the

Figure 2. Cartouche (detail) of *A Map of the Most Inhabited Part of Virginia,* by Joshua Frye and Thomas Jefferson's father, Peter Jefferson, 1775 version. (Monticello/Thomas Jefferson Foundation, Inc. Photograph by Edward Owen)

"post-revolutionary youth." On the one hand, he complained about those members of the next generation who did not want to reform the institution at all but were merely interested in maintaining the status quo. On the other hand, he censured those who in his opinion were, as was Coles, too precipitate and passionate in their wishes to end slavery. Thus, Jefferson condemned the nineteenth-century polarization toward a reactionary proslavery position, on one side, and the radicalization of immediatist abolitionism, on the other. Within the logic of his argument, his own generation represented the progressive middle between an unprogressive older generation and a regressive younger one. In his view, only a gradualist attitude to slavery could occupy the rational middle ground between too little and too much emancipation, between too late and too soon a reform.[92]

In the letter to Coles, Jefferson chose a remarkable image for the precarious instability of this middle position. Quoting from Virgil's *Aeneid,* he identified himself with "old Priam" during the fall of Troy, too weak to fight but nevertheless (courageously or foolishly) taking up the armor of his dead son Hector, only to meet a cruel death by the hand of Achilles's son.[93] Among the protagonists of this part of the myth, the war between two nations is also a fatal combat between two generations, an intergenerational chaos of fathers killing sons, and sons fathers: Priam's son Hector is killed by Achilles, whose son Pyrrhus kills Hector's father, Priam, while Pyrrhus's father, Achilles, is killed by Priam's son Paris. If Jefferson was only half-conscious of the quasi-parricidal context of this citation, it indicates how seriously he took the generational conflict over different modes of time perception in general and the timing of antislavery reform in particular.

As in the opposition between American masters and slaves, Jefferson imagined the opposition between two generations as a potentially warlike antagonism between two nations.[94] The comparison between "bidding old Priam to buckle the armor of Hector"[95] and Coles's attempt to recruit Jefferson for the antislavery cause may initially seem merely a self-deprecating reference to the weakness of old age. Yet between the lines, Jefferson strongly criticized Coles for trying to engage a venerable patriarch in a suicidal combat that would be unable to prevent the fall of Troy, or solve the slavery problem. The fact that Priam's heroic son Hector is already dead here–the absence of the prototypical "good son"–could be read as a further expression of Jefferson's disappointment with the younger generation. After all, if he had merely wished to use his age as an excuse for his passivity, rather than to combine it with a severe moral reproach, he might have found a more benign example within the eighteenth-century cult of patriarchal figures.[96]

Jefferson thus also directed his criticism of an unprogressive time consciousness to a group whose difference was not defined by the modern categories of race, class, or gender, but simply consisted in age and political opinion. As the letter to Coles illustrates, the limits of rational time could be found even within the group with whom Jefferson might be expected to have identified most: male republican citizens of Virginia. Also among relative social equals, he illustrated the opposition between different temporalities by use of belligerent metaphors, imagining it as an existential conflict about the power to define America's future. In the slavery question, if Jefferson's defensive-offensive language is to be read literally here, a rational time perception almost seemed a reason to fight (civil) wars. In a less literal reading, it still constituted a crucial factor in his vision of American progress.

Thus, Jefferson's reflections on divergent personal temporalities served different social and political needs. Excluding from America's future European monarchs and American slaves alike, he sought, first of all, to devalue two institutions that he identified with an outdated *ancien régime:* established aristocracy and chattel slavery. In his identification of aristocratic women with an unprogressive lifestyle, essentialist prejudice was relatively limited. Jefferson seldom spoke of "women" as a group, considering their lack of rationality to be chiefly the result of social position, education, and habits. Thus, he hoped that the education of his daughters, for instance, would enable them to act, in principle, as "rationally" as men (if only in their proper sphere). When he criticized Paris women, and not men, for their lack of temporal restraint, it was mainly because he could depict them as the embodiment of the entire *ancien régime.* Although this gendered criticism of the Old World would not have been possible without an emerging essentialism in his culture, Jefferson cannot generally be said to have assumed an inborn intellectual inferiority of women.

By comparison, his depiction of black slaves as creatures "naturally" incapable of foresight was his most problematic exclusion from a rational time perception, especially as it has turned out to be quotable by later racists. Of all the social and political conflicts, real or imagined, between people of different "temporal orientations," it was probably the possibility of an open struggle between black slaves and white masters that most frightened Jefferson. It came closest to contemporary constructions of an existential combat between master and bondsman by all standards, obliging him to fight on the wrong side and perfectly combining the dangers of a "class" with those of a "race" war. In some contexts, especially in his calls for colonization, it made more sense for Jefferson to emphasize an environmentalist perspective and imply that the slaves lacked the attribute of foresight due to their dependent position in slavery. More often, however, the difficulties of his argument for gradual abolition inclined him to suggest "natural" inferiority by criticizing, for instance, a lack of literary restraint not only among black slaves in America but also among free blacks in Europe, such as Ignatius Sancho. More obsessively than with any other group of people, Jefferson dwelled on the possibility that the supposed deficiency might not be due to cultural factors but could be "fixed in nature" and thus essentially, innately, and unalterably tied to their "colour."

What still distinguished Jefferson from modern racists was his own attachment to rational time in the context of his Newtonianism. It was essential to his view of himself as a future-oriented man of the Enlightenment that he could never be dogmatic on the point of innate black inferiority. Lest he, like the disappointing Romantics among the "post-revolutionary youth," should appear present-centered himself, he could not permit himself to do more than leave open to future observation the crucial question of nature and nurture.[97] Moreover, he still understood the species of irrationality he perceived among blacks as a universal phenomenon, not predominantly tied to inferior "black" blood, as later racists would claim, but typically occurring in the most blue-blooded families in Europe and generally threatening to happen to everyone under certain conditions. Finally, it even seemed doubtful that an exclusive reliance on rational time was the surest way to happiness. As I discuss in the next chapter, there were moments when Jefferson wondered whether his paternal punctuality was in fact an unsatisfactory mode of time perception, compared with the unique pleasures of enjoying past and present moments for their own sake.

Chapter 3

Sentimental Time

American and European writings from the second half of the eighteenth century abound with expressions of skepticism about the practical and theoretical significance attributed to rational clock time. In particular, late Enlightenment writers questioned the connection men such as Jefferson liked to establish between their privileged access to mechanical time measurement and their identity as forward-looking fathers or masters taking responsibility for others. In a novel that Jefferson included in his book catalog, Diderot's *Jacques le fataliste et son maître* (written 1773–75),[1] the "maître" is depicted as a foolish creature entirely dependent on "les trois grandes ressources de sa vie," namely, his watch, his snuffbox, and his servant Jacques. A caricature of eighteenth-century aristocratic life, the master does nothing but look at his timekeeper, take snuff, and call on his servant for help or entertainment.[2] Contrary to the claims to self-determination and social control inherent in Jefferson's rational time, in other words, the identity of the master consists in three things: his passive dependence on his servant's narratives and activities, his continual consumption of the product of colonial slave labor, and his useless fixation on the status symbol of his watch. Unfortunately for the master, it never really matters what time it is in Diderot's novel, whose antilinear structure makes the knowledge of clock time conspicuously irrelevant.

In its structural ridicule of rational time, its ironic clock symbolism, and its satirical inversion of the master-servant relationship, *Jacques le fataliste* took its cue from Sterne's *Life and Opinions of Tristram Shandy,* which Jefferson owned in various editions.[3] Containing in its first pages the most famous ridicule of clock time in eighteenth-century literature, Sterne's novel presents a mode of discussing and overcoming the linear order of a rational time perception. As the narrator Tristram points out at the very beginning, he "should have made a quite different figure in the world" if his parents had not organized their lives by an exaggerated attention to clock time. His father, Walter Shandy, personifies the questionable aspects of paternal punctuality. Tristram describes him not only as "one of the most regular men . . . that ever lived" but as the opposite of a self-determined *paterfamilias,* being "in truth a slave" to his "extreme exactness." Walter's regular winding of the "large house-clock" and the precise timing even of such "little family concernments" as his marital duties are the first of a long train of causes and effects that eventually lead, via "an unhappy association of ideas which have no connection in nature" in the Lockean mind of Mrs. Shandy, to Tristram's "misfortunes."[4]

Thus, *Tristram Shandy* turned the claims to social control and foresight upside down: it presented Walter Shandy's extreme punctuality as the cause of essentially unnatural effects. Moreover, it demonstrated that his slavish obedience to clock time was incapable of providing for the welfare of future generations (which for Jefferson in serious contexts would have been the ultimate moral reproach).[5] In exploring the full implications of Locke's *Essay Concerning Human Understanding,* Sterne's novel questioned the assumptions of the universality, linearity, and homogeneity of time, discussing the consequences of two divergent temporalities. Apart from the externally measurable progression of rational time, as described in the previous two chapters, a subjective concept of time gained prominence in the writings of Sterne, his admirer Jefferson, and many others. In a sense, this internalized *sentimental time* was a continuation of older approaches to the human condition. Like these older approaches, it focused less on the permanence of natural laws than on the theme of *vanitas.* Sentimental time was the time of human perceptions, feelings, and actions, a time structured by the aesthetic experience of irretrievable moments rather than by the predictive aims of the natural sciences. When writers in the second half of the eighteenth century questioned an exaggerated faith in universal progress, they often did so by stressing these human aspects of time perception. The late Enlightenment sense of a discrepancy between a rational and a sentimental time perception was at the root of the collective experience of an "acceleration" of time, which became relevant for Jefferson in both his personal life and his historical experience.

Despite growing tensions between rational and sentimental time in the crises of the late eighteenth century, however, Jefferson tended to integrate both approaches into his worldview, within a gradualist conception change. Instead of understanding rational and sentimental time as mutually exclusive, he usually allowed them to gain more or less relevance in his arguments, depending on the context. Thus, his recommendations of a "rational life" and his social exclusions from a rational time perception were not simply binaries by which he sought to consolidate his position through the exclusion of some inferior "other." Much as Jefferson feared a proto-Hegelian inversion of roles as the result of a present-centered lifestyle, he likewise feared that an *excess* of rational future-orientation might bring about an inversion of roles. Like many of his contemporaries, he thought that an exclusive reliance on rational time might eventually cause melancholy and disease, feelings of alienation and passivity. This assumption was part of a tendency in the late Enlightenment to question the validity of a rational future-orientation, from the viewpoints of both epistemology and practical morality. As David Hume saw it, knowledge of the future was not founded in reason, but merely in "custom."[6] And for Kant, while future-orientation was the faculty that made human progress possible, it was also what made men most miserable–as he argued in an essay dealing with the philosophy of the most radical eighteenth-century critic of rational time, the watchmaker's son Jean-Jacques Rousseau.[7] Even Adam Smith, today best known for a time-saving division of labor, criticized an exaggerated attention to the precision of clock time.[8]

Thus, while Jefferson denounced various groups–aristocrats, slaves, and the "post-revolutionary youth"–for their present-centered lifestyles, he showed himself in other contexts rather inclined to enjoy the sentimental quality of time perception, regardless of its significance for future progress. Especially in private letters to family and friends, Jefferson dilated on the sentimental value of unique moments in his past and present life. On these occasions, he sought to emulate the literary preferences of his early years–Edward Young, James Macpherson's Ossian, and especially Laurence Sterne. In their imagined company, Jefferson was able to introduce a sentimentalized version of the *vanitas* motif into his writings. Moreover, he entered into the ambivalent world of what has been described as the "great discovery of the eighteenth century," the double-edged experience of memory.[9]

"Time wastes too fast"

Jefferson's earliest preserved letter to a friend contains a humorous questioning of the universality of clock and calendar time.[10] Writing to John Page on Christmas Day, 1762, a youthful Jefferson, who had just finished his studies at William and Mary, related how on a day that was to all others "the day of greatest mirth and jollity" he had been overwhelmed by a series of misfortunes, indeed "more and greater misfortunes than have befallen a descendant of Adam for these thousand years past . . . and perhaps, after excepting Job, since the creation of the world."[11] Doubly distinguished in his time perception from the rest of human society–from everyone in the present celebrating Christmas (the beginning of the Christian calendar) and from almost everyone in the past since Adam and Eve (presumably the beginning of time itself)–Jefferson considered himself the object of a satanic conspiracy aiming at his destruction, or at least at the destruction of his most cherished property.[12] Having discovered that his pocketbook, silk garters, and music sheets had been eaten up by rats during the night, he found that his watch had been damaged as well:

> When I went to bed, I laid my watch in the usual place, and going to take her up after I arose this morning, I found her in the same place, it's true! but *Quantum mutatus ab illo!* all afloat in water, let in at a leak in the roof of the house, and as silent and still as the rats that had eat my pocket-book. . . . It's my opinion that the Devil came and bored the hole over it on purpose. Well, as I was saying, my poor watch had lost her speech. I should not have cared much for this, but something worse attended it; the subtle particles of the water with which the case was filled, had, by their penetration, so overcome the cohesion of the particles of paper, of which my dear picture and watch-paper were composed, that, in attempting to take them out to dry them, good God! *Mens horret referre!* My cursed fingers gave them such a rent, as I fear I never shall get over.[13]

This image of Jefferson's watch "all afloat in water" can be seen to epitomize the sentimental opposition to rational clock time that would come to characterize many of his writings. In his later linguistic repertory, metaphors related to water in the context of a sublime nature–waves, floods, torrents, or oceans–would play an important role in descriptions of his complex experiences of time and history. As an old man Jefferson would occasionally leave the enlightened control of clock metaphors behind altogether and speak, in almost Romantic terms, of an immeasurable "ocean of time."[14] In retrospect, therefore, his early anecdote describing how the mechanical order of a watch was dissolved by water seems to have anticipated this imagery: over the course of the following sixty years, the rivulet of a small leak in the roof flooding his watch would eventually flow into the sea of a Romantic temporality.

In the adolescent context of the letter, reflections on the literary expression of time philosophy were hardly what was most on Jefferson's mind. He was more concerned, on this occasion, with expressing his dissatisfaction with having to read "the old dull scoundrel," Edward Coke,[15] and most of all with having to leave the young women of Williamsburg. While he also expressed his rebellion against the future-oriented discipline of his law reading, the main point of the text was to talk about his beginning courtship of Rebecca Burwell, whose "dear picture" had apparently been destroyed by the "penetration" of the water.[16] However, it is significant that Jefferson explicitly gave priority to the picture of a young woman rather than to the functioning of his watch ("I should not have cared much for this"). The image of the "poor" speechless watch unable to function because of an unforeseen intrusion of water directly opposed the universal order attributed to clock time. In contrast to the optimistic expectation of perpetual control and progress underlying the "clockwork" conception of rational time, the image of a stopped watch alluded to human finitude and death. Combined with Jefferson's youthful opposition to the rest of society, who were celebrating the beginning of the Christian calendar, while he had to deal with his flooded watch, the anecdote of the Page letter initiated a counterdiscourse to rational time in his writings.

Although in its full implications hardly Jefferson's intention, his metaphorical urge to water down the order of clock time was part of a philosophical tendency to reveal as an illusion the objective measurement of time that the increasingly accurate clock technology of the eighteenth century seemed to promise. It pointed toward the ironic clock symbolism questioning the human control of time that was developing in the contemporaneous writings of Diderot and Sterne on the other side of the Atlantic. Demonstrating the inadequacies of mechanical timepieces, the young Jefferson joined his contemporaries in emphasizing that the human experience of temporality consisted in more than the simple measurement of identical points on a line extending toward the future. He wished to prolong and retain moments of lived experience for their own sake, to prevent the present from yielding to the future and becoming past itself. In this context, he did not idealize the future as the object of human improvement, but perceived it as a potential threat. It appeared linked, not to progress, but to deterioration, because it contained the loss of present happiness and, ultimately, the loss of life.

In his reflections on time, then, Jefferson not only dealt with death by referring to the long-term progress that posterity would enjoy. He also coped with human finitude by emphasizing the themes of loss, rupture, and separation. Locating a traditional theme of human *vanitas* in an eighteenth-century context, he replaced the conventional humility of *memento mori* by a greater stress on the psychological drama in the sentimental subject's present life. Although often expressed in religious language, as in Edward Young's *The Complaint; or Night-Thoughts on Life, Death, and Immortality* (1742–45), which the young Jefferson commonplaced extensively,[17] this secularization of *vanitas* tended to imply a certain amount of rebellion, an obstinate clinging to the present, culminating occasionally in the radical claim to prefer death altogether to the destruction of present happiness.[18] This emphasis was reflected in the tendency to take recourse also to ancient pagan writers, either from Greek and Roman antiquity[19] or from an imagined Celtic antiquity in the work attributed to a poet whom Jefferson's later correspondent Madame de Staël would call the "Homer of the North."

From early on, Jefferson was preoccupied with the topic of human mortality. As the handwriting, for instance, of his excerpts from Young's *Night-Thoughts* in his *Literary Commonplace Book* indicates, he started copying literary references to death and vanity while a schoolboy.[20] In the latter 1760s and the early 1770s he began to cultivate two literary penchants, one for the novels and sermons of Laurence Sterne, the other for James Macpherson's Ossian publications.[21] Jefferson praised Sterne's writings as "the best course of morality that ever was written"[22] and Ossian as "the greatest poet that has ever existed."[23] He clung to this estimate for decades, even though some of his contemporaries anticipated the nineteenth-century verdict on Sterne's novels as immoral and Ossian's ancient poetry was soon suspected not to be ancient or written by Ossian at all. In retrospect, it is difficult to reconcile this literary taste with the "typical" image of Jefferson as a didactic and forward-looking moralist. Young's nightly reflections on death, Sterne's sexually suggestive and ultimately also death-ridden narratives, and Ossian's obscure chants about a vanishing culture appear rather an incongruent selection of reading for a self-professedly "sanguine,"[24] optimistic, and progressive thinker. All three of Jefferson's literary preferences celebrated their opposition to a progressive temporality: in their different ways, Young, Ossian, and Sterne shared an attitude toward the passing of time that emphasized the destructive potential of the future.

Jefferson's well-documented enthusiasm for these writings demonstrates that there were moments when he needed a relief from the self-control of his "rational life," when the past and present did not appear mere preludes to a better future but became interesting in their own right, and when hopes for enlightened progress appeared impossible to maintain during existential experiences of loss and mourning. One such experience may have been the early death, in 1757, of his father, Peter Jefferson (whose first accurate map of the colony a proud son later included in his *Notes on the State of Virginia*). Another moment of Ossianic potential came with the sudden death of his close friend and brother-in-law Dabney Carr in 1773. For the inscription on Carr's tombstone at Monticello, Jefferson at first contemplated an adaptation of lines from Ossian, full of gloomy shadows, hidden moons, and mossy stones.[25] In a world where disease and premature death were a concrete presence, an abstract optimism must often have seemed particularly at odds with the more tangible experience of reality. The Ossian poems may therefore have appealed also to members of Jefferson's generation who had been brought up in a spirit of a future-oriented restraint. Their dark imagery and archaic heroism represented the antithesis of civilized control and order, while dressed in a sentimental language that Jefferson could find capable of expressing "the tender and the sublime emotions of the mind."[26] Through their anticlassical blend of genres, supposed temporal remoteness, and absence of a consistent argument, their appeal was mostly conveyed aesthetically rather than rationally–in a sense, more by "sensation" than by "reflection," to use Locke's terms. Their backward-orientation therefore never became a direct competitor to the restraint of Jefferson's educational writings. Instead, Ossian offered an ideal aesthetic supplement to the meager consolations an enlightened future-orientation had to offer.

Whereas Ossian tended to play a greater role in Jefferson's friendships with men,[27] it is arguable that Sterne's novels became more important in his relationships with women, in serious as well as in flirtatious contexts.[28] Probably about the time of Jefferson's marriage to the young widow Martha Wayles Skelton in 1772, he copied into his *Literary Commonplace Book* a passage from *Tristram Shandy* that condensed one of the central themes of the novel: the discrepancy between the two concepts of time that were implicit in Locke's concept of duration, an external rational time and an internal sentimental time. When in the summer of 1782 Martha Jefferson lay dying from complications of childbirth, after giving birth to six children in a decade of marriage that Jefferson later described as "ten years in unchequered happiness,"[29] she began copying her husband's commonplace version of Sterne: "Time wastes too fast: every letter I trace tells me with what rapidity life follows my pen. The days and hours of it are flying over our heads like clouds of windy day never to return more[–]every thing presses on–and every time I kiss thy hand to bid adieu, every absence which follows it, are preludes to that eternal separation which we are shortly to make!"[30]

From the farewell kisses onwards, the passage was completed in Jefferson's hand (figure 3). The couple thus adjusted Sterne's original to their purposes, making it read like a dialogue between a parting wife and husband. After Martha Jefferson's death, a grieving Thomas destroyed almost all objects related to her, but he kept this paper and a lock of her hair in a secret compartment of his desk. It was discovered there after his own death almost half a century later, having apparently been folded and unfolded many times.[31] The paper thus revealed its function as a sentimental souvenir in two ways: materially, the handwriting, combined with the conventional lock of hair, was supposed to preserve at least a part of Martha's physical reality; and ideally, the written lines sought to retain a shared moment of their lives, a psychological reality expressed by the adaptation of lines from *Tristram Shandy*. Husband and wife thus hoped to remain united in a dialogue that was both history and poetry, on the one hand materially preserved in ink on paper, historically documented, and adjusted to their particular situation, and on the other hand, ideally prolonging their shared existence by the reference to Sterne's fictional, timelessly significant reality.

Sterne's lines themselves foreground the eventual futility of the attempt to halt the flow of time by constructing a poetic keepsake. In this passage full of scriptural echoes in the ninth volume of *Tristram Shandy*, the narrator Tristram continues a theme that had become prominent in the beginning of the seventh volume: a race between the time of the narrator's life and the time of writing his life, in which his death increasingly threatens to "overtake" a suffering Tristram in the writing process.[32] In his hurried journey to escape death, the sentimental time of writing (based on subjectively varying processes in the mind) proceeds more slowly than the rational time of natural laws (observable in the physical processes of disease and aging). The passage conveys an alarming sense of acceleration, resulting from the "rapidity" with which the objective temporality of Tristram's (and Martha's) life "presses on" and "follows" the subjective temporality

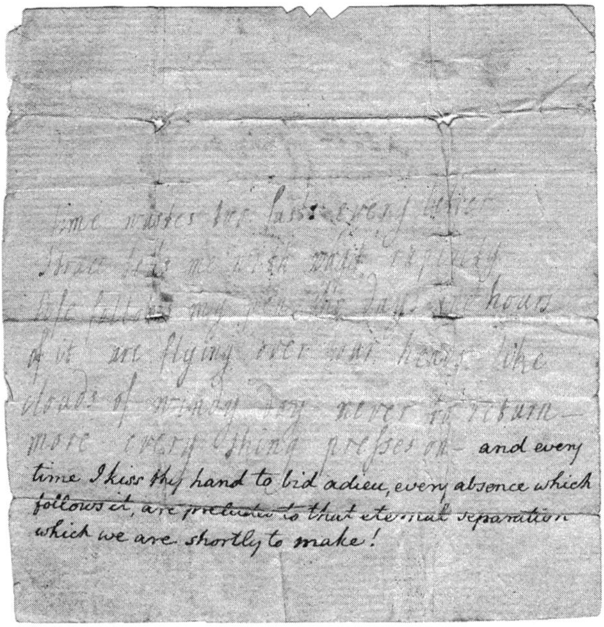

Time wastes too fast: every letter
I trace tells me with what rapidity
life follows my pen. The days and hours
of it are flying over our heads like
clouds of a windy day never to return
more everything presses on— and every
time I kiss thy hand to bid adieu, every absence which
follows it, are preludes to that eternal separation
which we are shortly to make!

Figure 3. Martha and Thomas Jefferson quoting from *Tristram Shandy,* 1782. (Courtesy of the James Monroe Museum & Memorial Library, Fredericksburg, Virginia)

of the "pen." This accelerated race with death becomes hopeless when the "days" of calendar time and the "hours" of clock time are compared to "light clouds of a windy day," an image that alludes to both biblical images for human vanity[33] and the movement of celestial bodies in space, the basis for the external measurement of clock and calendar time. These fast-moving clouds are "flying over" their heads: metaphorically, rational clock time is "overtaking" Tristram and his readers in the moment they are "tracing," writing or copying, the letters of the statement. The image for the acceleration of life thus already contains the moment of the approaching death itself.

In Sterne's original, the sense of doomed acceleration is even stronger when Tristram describes, in a line that is missing from Jefferson's commonplace version and from the dialogue version, how the hair of his "dear Jenny" is turning gray as they look at it.[34] By this overstatement referring to the rapidity of rational time, Sterne conversely illustrates the tragic slowness of sentimental time: since the universal temporality governing the physical process of aging cannot be presumed to have passed faster than usual, it has to be the particular temporality of their mental processes that is slowing down, unable to win the race of life and destined soon to be "overtaken" by death. Because of this slowness of human time perception, the sentimental subject tries in vain to cling to a present that is already becoming past. This irrational clinging to a passing moment results from the incongruity of the two temporalities in their Newtonian and Lockean world: Jenny, Tristram, and potentially all human beings tend to feel younger in sentimental time than they physically are in rational time, and tragically, they are capable of becoming aware of it. Closing with the uncharacteristically desperate (and at best half-ironical)[35] exclamation "Heaven have mercy upon us both!" the passage dramatizes the tragic dimension of the temporal human experience.

In a hopeless situation, Jefferson and his wife thus turned to a sentimental version of the *vanitas* theme in Sterne's novel, a secularized *vanitas* that focused less on a reunion in the next world than on the psychological drama of two lovers expecting their "eternal separation" in this world. Consciously or not, Jefferson omitted the allusion to Jenny's graying lock of hair, but he later supplemented the text with a lock of Martha's hair, in another necessarily futile attempt to transcend rational time and retain a link to a past presence. The drama of experiencing an acceleration of life due to the plural of two temporalities became crucial in various contexts of Jefferson's writings: apart from the tragedy of Martha's early death, also in the senseless "bustle" of modern European cities and in the rapid sequence of historical events in his political life. He dealt with the destabilizing phenomenon of acceleration by foregrounding rather than suppressing the experience of sentimental time in his writings. Martha and Thomas Jefferson appear to have received greater consolation from the fictional world of *Tristram Shandy* than, for instance, from the abstract forward-orientation of a purely theological argument. As in the case of Ossian, the aesthetically transmitted sense of sharing in a universal experience was a crucial supplement to more rational modes of explaining human mortality (which Jefferson would later employ, by contrast, in stoic references to his own approaching death).[36] Tristram's manipulative construction of his (written) lifetime offered the Jeffersons a welcome mode of aesthetically transcending the linear order of human finitude.

Martha Jefferson's death in September 1782 was, nevertheless, a traumatic experience for her husband, who passed several weeks in a delirious state. Indeed, Jefferson's grief may have exceeded the limits of what was considered usual or even decent for a man in his situation.[37] By his own account, Martha's death brought him close to wishing for his own death as well.[38] Looking back, he described her parting as a radical disruption of his "rational life": "Before that event my scheme of life had been determined. I had folded myself in the arms of retirement, and rested all prospects of future happiness on domestic & literary objects. A single event wiped away all my plans and left me a blank which I had not the spirits to fill up."[39] To conclude from the repeated term "event" in this short passage, especially from the expression "single event," which he also used in another letter,[40] Jefferson not only wanted to avoid more explicit references to Martha's death but appears to have been deeply shocked by its radical contingency. His experience of an isolated event capable of radically "wiping away" all his plans for the future was, in a sense, his personal Lisbon earthquake. It made his optimistic hopes for a progressive future appear ridiculous, exposing as an illusion the pursuit of "future happiness" in a self-determined "scheme of life." Moreover, it revealed the discontinuous aspects of the human experience and questioned the benevolence of the "design" of (human) nature, a nature that no longer formed a calculable continuity, but whose past was irretrievable, whose present too short, and whose future unpredictable.

The life-changing quality of a "single event" (rather than a predictable "course of human events") questioned the possibility of making plans on the basis of past experience, both on the level of personal lifetime and on the collective level of historical time. The shock of a catastrophic "single event" also reinforced a conception of the present as a unique moment that had to be experienced for its own sake. When Jefferson was recovering from his wife's death, his writings were characterized by a strong tendency to admit an open future into his life, especially in his exchanges with women: "Living from day to day, without a plan for four and twenty hours to come, I form no catalogue of impossible events. Laid up in port, for life, as I thought myself at one time, I am thrown out to sea, and an unknown one to me. By so slender a thread do all our plans of life hang!"[41] As in this letter to his friend and former landlady, Elizabeth House Trist, the Jefferson of the 1780s was able explicitly to deny, in his own life, the value of foresight that he was trying to impart to his daughters and nephews. This new mode of living in sentimental time–neglecting self-control and future-orientation–was not unlike the one he was criticizing in European aristocrats and Virginian slaves. After the contingency of his wife's death demonstrated that his control over the future had been an illusion, as his "scheme of life" had turned out so fragile that it could be wiped away in an instant, Jefferson admitted to having stopped focusing on order and necessity in human concerns. Instead of making plans for the future along the "slender thread" of rational time, he now claimed to consider nothing impossible in an unpredictable time to come.

To describe this openness toward human contingency, Jefferson used nautical metaphors that would become prominent in his later writings about American history.[42] Having been forced to leave the safe "port" of his marriage and of his first retirement, he felt cast into an uncontrollable element, the "sea" of a widower's life in politics. When Jefferson wrote these lines, he had not only figuratively but also literally crossed the Atlantic Ocean for what would be his only sojourn in the Old World. As he told another correspondent that year, his stay as American minister in Paris was closely related to his wife's death, as Congress would not have been able to induce him to end his retirement from public office otherwise: "The domestic loss I have sustained was the only circumstance which could have brought me to Europe."[43] Trying to come to terms with the disruption of his "scheme of life," he gave up some of his pretensions to control over the future by trying to view the remainder of his time as a journey through an "unknown" element to an uncertain destination. On this imagined voyage, he could accept the provisional character of his situation more easily. If there was no reliable method to achieve control over the future, as every day could take him to a new place and a new society, it was more legitimate to live exclusively in the limited situation that the present offered.

As his letters from this period indicate, Jefferson became more inclined also to see the positive aspects of this conception of life as a journey. Aided by a renewed interest in Sterne, now especially *A Sentimental Journey through France and Italy* (1768), he liked to style himself as a "sentimental traveller."[44] To be sure, most of Jefferson's European travel notes contained practical information for future use in America.[45] Yet some of his personal letters from this period presented scenes from his journey only as they influenced his own state of mind in the present. Like the reflections of Sterne's Yorick, they emphasized subjective processes in his mind at the expense of external descriptions of places. This internalization of his actual journey through space was paralleled by an internalization of his temporal journey through life. While traveling through southern France in 1787, for instance, he wrote to Madame de Tott that he had successfully accommodated himself to the "noise, dirt, and disorder" of his simple inn at Marseilles by

> philosophizing on the objects which occur. The plan of my journey, as well as of my life, being to take things by the smooth handle, few occur which have not something tolerable to offer me. . . . "A traveller, sais I, retired at night to his chamber in an Inn, all his effects contained in a simple trunk, all his cares circumscribed by the walls of his apartment, unknown to all, unheeded and undisturbed, writes, reads, thinks, sleeps, just in the moments when nature and the movements of the body and mind require. Charmed by the tranquillity of his little cell, he finds how few are our real wants, how cheap a thing is happiness, how expensive a one pride. He views with pity the wretched rich, whom the laws of the world have commited to the cumbrous trappings of rank: he sees him labouring through the journey of life like an ass oppressed under ingots of gold . . . gobbled up by harpies of various descriptions with which he has surrounded himself. These, and not himself, are it's real masters."[46]

According to this half-flirtatious, half-serious letter, Jefferson was able to extract enjoyable aspects from the unpleasant material "disorder" of his present situation. Actively transforming his experiences by reflective processes in the mind, he was able, as he also put it on other occasions, to "take things by the smooth handle."[47] He sought to enjoy the present moment for its own sake, having made a conscious decision to find life "tolerable" again, the "plan of his journey" being not to make too many plans that could be wiped away. By sharing with Madame de Tott a supposed interior monologue about the "happiness" of traveling alone, Jefferson presented himself as a mixture of a secularized monk (in "his little cell") reflecting on the vanity of riches and rank and a literate version of Rousseau's natural man, who, living without society and unable to reflect on the future, "unknown to all, unheeded and undisturbed, . . . thinks, sleeps, just in the moments when nature and the movements of the body and mind require."[48] Thus describing the exact opposite of his future-oriented "rational life," Jefferson fused a place on his spatial journey through France with a moment of his temporal journey through life: his lonesome chamber in Marseilles contained not only all his belongings ("all his effects contained in a simple trunk") but also "all his cares," which were "circumscribed by the walls of his apartment." His "little cell" was the spatial equivalent of the temporal extension of his thoughts, in a sentimental inner life that took place mainly in the present.

Jefferson's Sterne-inspired happiness of traveling from one moment to the next implied an interesting idea of liberty. Unlike the "wretched rich

. . . labouring through the journey of life like an ass oppressed," forced by "the laws of the world" to have others work for him, the traveler staying anonymously at a noisy and dirty inn could feel free from this sort of oppression. Living exclusively in the present, Jefferson implied that he was his own "master," an idea of mastership that depicted foresight and self-control as a burden rather than an asset, contrasting sharply with the concept of the *paterfamilias* in his family letters. In a sentimental correspondence with a French lady, Jefferson's sense of the relation between an identity as "master" and a future-oriented time management was directly inverted.[49]

A week before writing to Madame de Tott, by contrast, Jefferson had sent a serious letter to his daughter Martha, admonishing her to structure her life by an abstract temporal order, the only means of ensuring her "future happiness."[50] His contradictory position–teaching others what he was not inclined to do himself–might appear to be an instance of Jeffersonian "hypocrisy" in retrospect. However, his sentimental equation of happiness with a present-centered life was no direct contradiction to his exhortation to his teenage daughter. Both statements can be seen as consistent expressions of the same worldview: in contrast to his letter to Martha, Jefferson's expressions of yearning for a simple life in the present were, to some extent, humorous overstatements, owing to his scrupulous efforts to present himself to French society as an amusing mixture of an American "son of nature"[51] and a Shandean "hypochondriac Englishman."[52] Furthermore, rational and sentimental time management were not mutually exclusive absolutes in Jefferson's pursuit of happiness. Even if the letter to Madame de Tott had been entirely serious, the particular space and time of its writing made a difference for Jefferson: whereas he hoped that his daughter could look forward to her "future happiness," he saw himself as a middle-aged widower who, after the disruption of his personal plans, could at best pursue occasional moments of happiness in the present. Apart from this difference in years, there was also a difference of space, which would become crucial in Jefferson's later historical writings. While Jefferson envisioned Martha achieving happiness by a quantifying "distribution" of time in America, his letter to Madame de Tott reflecting a qualifying approach to time was supposed to become part of the conversation of a French salon.

As discussed in the previous chapter, Jefferson tended to connect the absence of rational time management to institutions that he claimed belonged to the Old World, such as established aristocracy and chattel slavery. In his later years, he would try to connect the unpredictability of sentimental time to the chaos of European history, opposing it to what he hoped would be the linear order of American progress. During his sentimental journey through the Old World, he considered it appropriate to adjust to a "European" temporality, whether in French conversations or in Sterne's novels. Supplying the experience of change with a greater emotive quality and aesthetic appeal, Jefferson's references to sentimental time were able to include both the pessimistic and the optimistic nuances of a secularized approach to human *vanitas.* When the darker side of sentimental time dominated his thoughts, he developed an attitude of skepticism toward the linear teleology of human progress, inducing him to lament the eventual futility of "human events." Yet when its lighter shades prevailed, Jefferson was able to express a heightened sensitivity to the uniqueness of the present, which he knew had to be enjoyed all the more, precisely because it was already becoming past.

Memory

When during his presidency Jefferson reminisced with an "ancient friend," Comte Diodati, about their shared experience in the diplomatic corps at Versailles two decades earlier, he idealized this time as "days of tranquillity and happiness." This phrase pointed to a contrast to the less tranquil and happy days that had followed: from Jefferson's hindsight perspective, his years as American minister in Paris from 1784 to 1789 had coincided with the twilight years of the *ancien régime* in France. This period before the collective experience of great historical chaos had been, as a sentimental Jefferson later remembered it, full of hopes for future progress that were soon to be "overtaken" by the historical "tempest" of the French Revolution.[53] As president, Jefferson wrote to another acquaintance from these years, the French *idéologue* Pierre-Georges Cabanis: "In those days how sanguine we were! and how soon were the virtuous hopes and confidence of every good man blasted! and how many excellent friends have we lost in your efforts toward self-government, *et cui bono?*"[54]

From the perspective of the early nineteenth century, the "days of tranquillity and happiness" that Jefferson had passed with European liberals in Paris—"then the seat of the most amiable and polished society of the world," as he later wrote to Madame de Staël[55]—seemed irretrievably lost. In a heterogeneous sentimental time, prerevolutionary France was different in quality and entirely absent from a present that was characterized by the loss of friends and the destruction of republican hopes. This past was no longer linked to the future by the possibility of rational foresight. As the

French Revolution had demonstrated, Jefferson's contemporaries in the 1780s had been unable to make valid predictions on the basis of their historical knowledge, and the virtuous beginnings of their republican hopes had resulted in the moral opposite–despotism, war, and terror. Thus, the retrospective vision of the French Revolution condensed the experience of rupture and "loss" of a vanishing past, a past so different in quality that it had to be remembered more for its own sake than for its didactic value. The fundamental political and cultural changes occasioned by the French Revolution might therefore appear today to have initiated a mode of remembering the past on its own terms, in a modern historical sense.

In Jefferson's case, however, the radical turn of the French Revolution was not the only cause, and probably not the most important one, for the sentimentalization of the past. Instead of seeing the French Revolution as a decisive rupture in world history, as became the dominant tendency on the Continent, Jefferson regarded 1789 as a predictable development and then refused for a notoriously long time to see anything particularly shocking in the Terror. In his later years, he even tended to deny that French history had any direct relevance for America's future. From his perspective as president, the significance of 1789 or 1793 appeared dwarfed by the much greater world-historical influence he associated with two American events, the achievement of independence in 1776 and his election to the presidency in the "Revolution of 1800."[56] Since France could at best claim third place among the nations possessed of Jefferson's loyalty (after Virginia and the United States),[57] his nostalgic musings on past "hours of tranquillity and happiness" in prerevolutionary France may have been due largely to his tendency to accommodate his language to that of his correspondents. Most importantly, the emotional contrasts that characterized Jefferson's comparisons between pre- and postrevolutionary France were strikingly similar to contrasts that he established in contexts that had nothing to do with the French Revolution.

Thus, the dualistic structure of Jefferson's memories of his time in France has to be interpreted in the larger context of eighteenth-century conceptions of memory. While the twilight years of the *ancien régime* in France offered him good opportunities to reminisce about a past period of his life, many other events occasioned a similar backward-orientation in his writings. In a sense, his sentimentalization of the past was a direct consequence of his attachment to the present: if, as discussed above, Jefferson was learning to understand the present as a moment whose uniqueness had to be enjoyed, the significance of the past can be expected to have changed for him as well. Not surprisingly, therefore, his middle and later writings abound with celebrations of bygone moments. Of course, this was due in part to his primary literary medium: especially at the beginning of personal letters, references to previous letters, previous meetings, or the general passing of time had almost a ritual character. In letters in which Jefferson gave more than the usual amount of attention to this topic, most often addressed to close friends or family, it is possible to discern a pattern in his recollections. This pattern owed to the epistemological problems of his empirical frame of mind.

In a Lockean worldview that tended to reject Platonic concepts of anamnesis, human memory was an especially complicated matter. Locke's *Essay Concerning Human Understanding* depicted memory, one of two modes of what he called "retention," as a particularly unreliable faculty. Not only was the mind initially a "white paper void of all characters,"[58] but it was destined to become "void" again if all perceptions ceased. In his efforts to counter the rhetoric of innatist arguments, Locke described the elusive character of human memory in poetic language resonant with allusions to human vanity:

> *Ideas* in the mind quickly fade and often vanish quite out of the understanding, leaving no more footsteps or remaining characters of themselves than shadows do flying over fields of corn; and the mind is as void as if they never had been there. . . . Thus the *ideas* as well as children of our youth often die before us; and our minds represent to us those tombs to which we are approaching: where, though the brass and marble remain, yet the inscriptions are effaced by time and the imagery moulders away. *The pictures drawn in our minds are laid in fading colours,* and if not sometimes refreshed, vanish and disappear.[59]

Before a human being died, many of his or her ideas had died already, having constantly been "fading" with the passing of time. In order to prevent their complete disappearance, they had to be brought back into "sight," to be "refreshed," "revived," and "painted anew" in the mind. Locke's emphasis on the sense of sight went back to ancient memorizing techniques of the "art" of memory, which used visual images (as from architecture) to memorize, for instance, different parts of a speech. For Locke, the best way to compensate for the natural decay of memory consisted in the "repetition" of these visual impressions, which was most effective if they were accompanied by "*pleasure* or *pain.*"[60] In the eighteenth century, David Hume followed Locke in emphasizing the emotive and sensory (especially visual) character of memory. In his *Treatise of Human Nature* (1739–40) he spoke of the "decay" of ideas in memory, which became "very weak and feeble" with time. He now positively distinguished the (recent) memory from the imagination by its greater "vivacity" and more "lively colours."[61]

Aspects of these conceptions of memory were not only reflected in the letters an older Jefferson wrote to his European friends, trying to recollect prerevolutionary happiness by emphasizing the contrast of past pleasures and recent pains. They also shaped his correspondence, for instance, with his daughters during his stressful experience of American politics during the 1790s. Martha's letters were to Jefferson like "gleams of light" cheering "a dreary scene" and bringing back the contrasting memory of her "fire side, where the single evening I passed at it was worth more than ages here."[62] As the shading of this imagery indicates, Jefferson hoped to refresh the warm tints of his memories by repeating them in the act of writing. The remembrance of cheerful scenes with Martha became the subject of an entire letter on another occasion:

> Having no particular subject for a letter, I find none more soothing to my mind than to indulge itself in expressions of the love I bear you, and the delight with which I recall the various scenes thro which we have passed together, in our wanderings over the world. These reveries alleviate the toils and inquietudes of my present situation, and leave me always impressed with the desire of being at home once more, and of exchanging labour, envy, and malice for ease, domestic occupation, and domestic love and society, where I may once more be happy with you, with Mr. Randolph, and dear little Anne, with whom even Socrates might ride on a stick without being ridiculous. Indeed it is with difficulty that my resolution will bear me through what yet lies between the present day and that which . . . my mind has determined to be the proper one for relinquishing my office. Tho not very distant it is not near enough for my wishes.[63]

In this letter, Jefferson emphasized the emotive quality of his memories, their qualification by pleasure and pain, which was necessary for his passive memory to "retain" and remember them. Like his reminiscences with European friends, his recollections of the past were conveyed by the contrast between conflicting emotions, between the "delight" in his "reveries" and the "toils and inquietudes" he was experiencing in the present, between "labour, envy, and malice" in the present and "ease, domestic occupation, and domestic love and society" in a past that he hoped would be repeated in the future. By the temporal construction of his composition–moving from a pleasant past to an unpleasant present, and from this unpleasant present "back" to a pleasant future–Jefferson's letter illustrates the potentially circular character of his concept of memory. By repeating and reviving a past moment in the act of remembering, he hoped to "impress" his mind to effect an actual repetition in the future of a remembered state of happiness. Instead of the self-control necessary for the linear forward-orientation of rational time, the cyclical movement of sentimental time had a "soothing" function: the repetitive "reveries" of his past family idyll–a mode of recollecting the past first by dreaming, then by writing, and finally by actually repeating it in the future–were supposed to "alleviate" the stress of his life in politics. In the same spirit, Jefferson's Heart addressed the Head in 1786: "Oh! my dear friend, how you have revived me by recalling to my mind the transactions of the day! How well I remember them all, & that when I came home at night & looked back to the morning, it seemed to have been a month agone. Go on then, like a kind comforter & paint to me the day we went to St. Germains. How beautiful was every object!"[64]

Similarly to Jefferson's writings to Martha, these lines to Maria Cosway demonstrate the "reviving" and "comforting" function of sentimental memories, their ability to help restore a past state of happiness in the present. By its own account, the Heart already remembered the events it had experienced in Maria Cosway's company ("How well I remember them all") but was nevertheless "revived" by the Head's narrative repetition of them. The therapeutic powers of Jefferson's memories were based on the repetition made possible by a subjective time perception. As in *Tristram Shandy,* the narrator's mind, rather than external laws of nature, provided the temporal order of the text. When Jefferson's Heart recollected a day with Cosway in the evening, accordingly, a whole month seemed to have passed in a qualifying time perception measured by the "mass of happiness" of lived experience. A precious "single evening" at Martha's fireside could likewise be "worth more than ages" in politics, and the "few hours" Jefferson spent with his friend Chastellux reading Ossian at Monticello could pass as quickly as the same number of "minutes."[65]

The subjectivity of sentimental time also had a destructive dimension, however. Once it was emancipated from the linear order of rational time, the experience of a single "hour" of human misery could be worse than "ages" of political oppression, as Jefferson put it in a rare instance of sympathy with American slaves.[66] When he experienced the present as a "dreary scene" characterized by "toils and inquietudes," he likewise suffered from the potential slowness of an internalized temporality. As he wrote to his daughter about the moment of their reunion, "Tho not very distant it is not near enough for my wishes." Attempts to retain past moments by narrative repetition at best "alleviated" his stress temporarily but could not entirely reconcile him to a painful period in the present. Sentimental recollections were intrinsically ambivalent: their appeal was always based on the emotional contrast to a less satisfactory present or future. As Jefferson realized, his efforts to regard life as a journey from one moment to the next, to experience and remember each moment for its own sake, were ultimately as imperfect a mode of achieving happiness as the contrary extreme of an exclusive reliance on a "rational life."

This insight decisively shaped Jefferson's moral advice to other men. He repeatedly admonished them to abstain, if possible, from sentimental journeys to Europe, especially from entanglements with European women. The ambivalence of memories from the Old World was especially dangerous for young, impressionable men, but Jefferson admitted that it could even do harm to "men of more sober age" like himself: "They are subject ever after to recollections mixed with regret, their affections weakened by being extended over more objects, & they learn new habits which cannot be gratified when they return home. Young men who travel are exposed to all these inconveniences in a higher degree. . . . Their eyes are forever turned back to the object they have lost, & it's recollection poisons the residue of their lives."[67]

Sentimental recollections could "poison" life if they were from a reality that was "lost" and could not be repeated in the future.[68] Anticipating the international theme of later American novels, Jefferson's attitude toward young Americans traveling to Europe epitomized the problem of "poisonous" recollections and "eyes forever turned back" toward a lost past, as he thought that American travelers could not, or should not, transfer their Old World experiences to the New World (at least in morals and politics). Jefferson considered memories from European travels dangerous for young men both on a public and on a private level. He feared that the "glare of pomp & pleasure" and the "spirit of female intrigue" at the courts could corrupt young Americans politically, while their personal relationships would be spoilt by recollections of "the voluptuary dress and arts of the European women."[69] As discussed above, Jefferson saw a close connection between the Old World and a present-centered or backward-oriented temporality, in both its charming and its destructive aspects. He was ultimately as ambivalent toward sentimental journeys through European society as he was toward sentimental journeys through life: due to the stark emotional contrasts and the circular structure of his conception of memory, remembering the past for its own sake could be as disastrous to one's personal happiness as unprogressive European monarchies were to the happiness of mankind.

Especially during his final retirement, Jefferson's recollections of the past became problematic for yet another reason, which again had its root in Lockean epistemology: the increasing fallibility of the human memory with time. As he wrote in 1814:

> Nothing is so painful to me as appeals to my memory on the subject of past transactions. From 1775 to 1809, my life was an unremitting course of public transactions, so numerous, so multifarious, and so diversified by places and persons, that, like the figures of a magic lantern, their succession was with a rapidity that scarcely gave time for fixed impressions. Add to this the decay of memory consequent on advancing years, and it will not be deemed wonderful that I should be a stranger as it were even to my own transactions. Of some indeed I retain recollections of the particular, as well as general circumstances; of others a strong impression of the general fact, with an oblivion of particulars; but of a great mass, not a trace either of general or particular remains in my mind.[70]

This passage is remarkable in the context of a Lockean conception of memory because it combines references to two different chapters from book 2 of Locke's *Essay Concerning Human Understanding,* the chapter on "retention" and memory and the chapter on "duration" and time. Whether or not Jefferson was actually thinking of the *Essay* when he wrote the letter, he likewise spoke of the "decay" of a memory that was unable to retain even "a trace" of his former recollections. As a result, he felt like a "stranger" to some of his own former actions, as if he were approaching Locke's decayed tombstones with inscriptions so defaced by time that it was impossible to know the names of the people they had once represented.

For Jefferson, the natural decay of his aging memory was not the only cause of his estrangement from his past. His impressions were prevented from "fixing" themselves in the memory also due to the particular "rapidity" of political change that he had experienced. To illustrate this rapidity, Jefferson used a simile familiar from Locke's *Essay* that he may also have remembered due to its ironic paraphrase in *Tristram Shandy.* Referring to the "figures of a magic lantern," he alluded to Locke's cyclical image of a "lantern, turned round by the heat of a candle" for his internalized concept of time gained by the "succession of ideas" in the mind. In the long eighteenth century that had followed the publication of the *Essay Concerning Human Understanding* in 1690, the implications of Locke's arguments had come to be subject to critical scrutiny from various angles. As in *Tristram Shandy,* the Lockean assumption of a subjective time perception had clashed increasingly with the Lockean assumption that this internalized time perception was somehow "constant and regular."[71] In 1814 Jefferson still referred to Locke's magic lantern, but he presented an accelerated version of it: the succession of ideas in his historical experience had taken place "with a rapidity that scarcely gave time for fixed impressions." The subjectivist implications of Locke's concept of time had been radicalized in Jefferson's letter, and its potential of acceleration now directly opposed Locke's assumption that there were "certain bounds to the quickness and slowness" of the succession of ideas.[72] Jefferson eventually referred to the work of his memory more radically than Locke, as an accelerated movement of the lantern, an absurd circle of blurred impressions that made lasting memories, unless written down and printed, utterly impossible.

In Jefferson's old age, the work of his memory was more important to him than ever before, as most of his life now appeared to consist in recollections. However, the significance of these memories was highly dubious: their accuracy often appeared as unreliable as their therapeutic powers appeared questionable. This uncertain perception of his past made it plausible for Jefferson to liken himself to an old watch slowly expiring: "Our machines have now been running for 70. or 80. years, and we must expect that, worn as they are, here a pivot, there a wheel, now a pinion, next a spring, will be giving way: and however we may tinker them up for a while, all will at length surcease motion. Our watches, with works of brass and steel, wear out within that period."[73] Writing to John Adams a few months after he had described himself as a "stranger" to his past, Jefferson recurred to the ironic clock metaphor of sentimental time, which he had been developing since his college days. It offered him a relaxed way of referring not only to the gradual decline of his bodily and mental faculties but also to his conflicting views of time. The image of an old watch fused the order of rational time and the idiosyncrasies of sentimental time, the quantifying and the qualifying aspects of human time perception. As I shall argue in the remainder of this chapter, this combination was Jefferson's dominant mode of dealing with the two conflicting "times" that structured his Newtonian and Lockean world.

Gradualism

As Jefferson's ironic clock metaphors illustrate, his tendency to question the universal significance of clock time pointed toward, but never quite equaled, the radical criticism of the next generations, which would be condensed in the diabolic clock metaphors of Edgar Allan Poe and Charles Baudelaire.[74] In contrast to the uncompromising Romantic "sons" like Poe (who was among the first students of the Jefferson-fathered University of Virginia), Jefferson was able to incorporate different approaches to time into his worldview, refusing to think of them as mutually exclusive absolutes. Accordingly, his writings did not evince the radical Romantic urge to abandon or transcend rational clock time altogether, by moving "out of time" or "above time," in Poe's and Emerson's words.[75] Instead, Jefferson was able to recognize the complexity of the experience of time without disavowing human longings for timelessness.

This flexibility was facilitated by the approximative character of Jefferson's Newtonianism. Instead of imagining an all-encompassing historical time that synthesized linearity and contingency, Jefferson allowed for different possibilities, an externally flowing time and a time that depended on potentially varying operations within the mind. At bottom, this flexible dualism was part of a transitional worldview that was as yet imperfectly secularized: Jefferson's rational time perception still sought to approach an everlasting version of God's eternity, and his sentimental time psychologized a traditional approach to human *vanitas.* Like many of his contemporaries, he sought to place himself in the middle ground between the two approaches. This middle ground lost relevance in the succeeding generations. In retrospect, Jefferson's two divergent temporalities may thus appear as a contradictory anticipation of modern modes of differentiation: his secularized approximation to eternity might be seen to have been radicalized in the natural sciences of the nineteenth century, in an industrialized economy, in visions of democratic progress, or in the coordination of public life generally. Alternatively, the heirs of his secularized version of a "merely" human temporality could be found in an increasingly professionalized realm of original art, in neorepublican narratives of political corruption, or in a separate sphere of "free time" and private life. Jefferson's own priority, however, was the compromise of a "mean time" between the two temporalities. Romantic dichotomies like those of public and private, science and art, were not fully developed in his thought, and even on occasions when their boundaries became apparent, he mostly sought to reconcile them within a worldview that cherished a version of the golden mean as the desirable mode of living and thinking.

It may thus be surmised that Jefferson did not feel much of a conflict when, for instance, he conscientiously strove for a future-oriented "rational life" in his meticulous record keeping, while simultaneously professing to enjoy the careless and spontaneous experience of the present. Like many of his contemporaries, he tended to think that both methods of organizing one's time had their benefits when applied jointly and with moderation but could become dangerous if applied exclusively or excessively. As has been analyzed in the context of Jefferson's medical reading, he agreed with contemporary studies in considering one-sided intellectual habits likely to cause disease and unhappiness.[76] In a similar spirit, he did not expect the future-oriented restraint of his rational time management to be capable, by itself, of guaranteeing tranquillity or felicity in every context. Much as he criticized the absence of rational time perception among European aristocrats, African slaves, or the "post-revolutionary youth," he did not consistently consider a rational "distribution" of one's time as the best road to happiness. There were situations when reason had to be complemented by sentiment and when rational calculations of future happiness proved unsatisfactory without the aesthetic enjoyment of unique moments in the present and past.

Jefferson's combination of rational and sentimental time not only helped him organize his personal life. It also provided solutions to more comprehensive problems in his conception of change. When Jefferson thought about change, he usually tried to imagine it, not as an uncontrollable movement in an undivided flow of time from an unknown past to an undefined future, but as a gradation of separate steps or "stages" leading from a fixed beginning to a fixed aim. Such gradual forms of change can be found in his writings on topics as diverse as individual education, universal history, generational rights and duties, the abolition of slavery, political and scientific progress, the process of aging, the progression of a disease, creative processes in the arts, the process of reading, and the progress of an enlightened conversation. In each case, change gained its order from the universal teleology of rational time, while each separate step or stage was characterized by a degree of individuality in sentimental time. Jefferson's gradualism thus managed at once to tame the contingency and to qualify the linearity of temporal change.

To some extent, education was the prototype of all other temporal gradations in Jefferson's thought. The points of beginning and ending were fixed (by birth and death), and the "mean time" between them was made up of different periods, imagined as separate, ascending or descending steps or stages. Within this gradualism, each cumulative rather than dialectical step was "separate and equal": Jefferson set a different standard for each period, as he attached different functions to each time of life, while maintaining that each period was equally important for the attainment of a set future aim. He stressed the importance of early impressions, of childhood and youth, for forming habits and moral dispositions.[77] On this foundation, parallel to the collective stadialism in contemporary works of universal history, he built up cumulative stages of reading.[78] This education by degrees was continued by his conception of gradually evolving political careers and the eventual decay of mental and physical faculties, which he expected to proceed in "soft gradations" until death.[79]

While unpredictable events were not completely absent from this gradualism, as they could come to characterize individual stages of development, they were rarely expected to transform its long-term future. In most cases, Jefferson tended to regard the structure of gradual rather than sudden change as "natural"–as the temporal equivalent to the spatial gradations in the "great chain of being" of natural history[80]–and therefore acceptable both as fact and as value. Hence, his frequently used *gradually, by degrees, pari passu, step by step,* and *in the mean time* could also be positively connoted expressions. They enabled him to deal with unpleasant situations in the present and the near future more easily, as they described them as periods that were only provisional stages in a long line of gradations–single steps in a progression of many other steps that would eventually lead, however unsatisfactory they appeared in the short term, to a more fulfilling long-term future. In pessimistic contexts, conversely, they could also make it possible for him to view the present as a "middle stage" whose progressive potential threatened to be capped, at some future point, by a one-sided modernization of the American republic.[81]

Like its intellectual predecessor in traditional Christian teleology, Jefferson's gradualism thus often resulted in a relativistic attitude toward the human experience of the present and near future. This semisecularized relativism was crucial in many different contexts of his writings, personal as well as political. It was part of a mentality that allowed conflicts of conviction or action to remain unresolved in a period that Jefferson regarded merely as "mean time," an approximative time that could be expected to be terminated sooner or later. The exact moment when the problem would be solved could not be known, but the eventual solution itself was rarely in question. As Jefferson wrote, for instance, in his humorous discussion with Anne Bingham about women's lives in Europe: "If we do not concur this year, we shall in the next, or if not then, in a year or two more."[82] Discussing more seriously the decisions for professorships in the future University of Virginia, he told a correspondent in 1814: "I agree . . . that a professorship of Theology should have no place in our institution. But we cannot always do what is absolutely best. Those with whom we act, entertaining different views, have the power and the right of carrying them to practice. Truth advances, and error recedes, step by step only; and to do to our fellow men the most good in our power, we must lead where we can, follow where we cannot, and still go with them, watching always the favourable moment for helping them to another step."[83]

This statement implied a combination of arrogance and humility typical of Jefferson's natural aristocracy: he situated himself and his correspondent midway between, on the one hand, an aristocratic impulse to "lead" and "help" the ignorant based on a superior knowledge of "truth" and, on the other hand, a democratic recognition of the limits defined by "the power and the right" legitimately claimed by those ignorant contemporaries. Analogously to this golden mean in the social and political hierarchy, Jefferson regarded the present situation merely as one step among others in the gradual disappearance of error and progress of truth. In such a spatial and temporal middle position, he advised pragmatic moderation and tolerance, implying that doing what was relatively best might sometimes be preferable to "always" doing what was "absolutely best."[84]

To modern eyes, Jefferson's lack of dogmatism may appear attractive in the context of theology. In other contexts, however, arguments that were constructed in gradualist terms very similar to those in the above letter–advocating "step by step" measures, accepting the provisional character of the present, and respecting the views of less enlightened citizens–now strike us as highly problematic in their connection to Jefferson's greatest moral problem: his increasingly evasive attitude toward the slavery question. In his 1814 letter to Edward Coles, for instance, Jefferson used a gradualist argument to advise Coles not to follow his inclination to emancipate his slaves at once but to wait until a more favorable moment in the future. He tried to persuade the younger man not to act rashly and produce a "shock" but to wait and "insinuate & inculcate" antislavery thought "softly but steadily, through the medium of writing and conversation" to convince Virginian citizens. The peaceful means of enlightened discourse were preferable to what Jefferson depicted as a violent rupture that was senseless and arbitrary. Since emancipation would only be a matter of time, Coles had to bring himself, "in the mean time" of the present, to subdue his passions. He had to refrain from "abandoning" both his property and, with it, his "country," because "the hour of emancipation is advancing, in the march of time. It will come; and whether brought on by the generous energy of our own minds, or by the bloody process of St Domingo . . . is a leaf of our history not yet turned over."[85] Since the aim itself was fixed, and only its historical conditions were in question, Coles, according to Jefferson, had to act cautiously in order to prevent a war between free blacks and Virginia citizens. Justified by a higher law, such a war threatened to end, Jefferson was afraid, in the "extirpation" of the masters.[86]

Like many of his European and Euro-American contemporaries, as has been said, Jefferson always argued for the gradual, rather than the immediate, abolition of slavery. During the last third of his life, in a kind of double gradualism, he even postponed the beginning of gradualist emancipation measures into an indeterminate future. He argued for a temporary maintenance of slavery in the "mean time" of the present, claiming that the public mind had to be gradually prepared to become "ripe" for the measure, that "ill-timed efforts" would do more harm than good, and that "time, which outlives all things," would in the long run "outlive this evil also."[87] Moreover, he used his gradualist concept of aging, as well as his gradualist theories of political generations, to explain his own silence in public discussions of slave emancipation. Taking into account that Jefferson's personal livelihood (including a considerable degree of luxury) always depended on the exploitation of slave labor, this gradualism tends to appear, in retrospect, like a poorly disguised mode of temporizing. It seems to present a hypocritical Jefferson who tactically claimed to postpone the problem in order really to eliminate it insofar as it concerned him personally, officially professing to hope for a future abolition of slavery, while secretly hoping that it would not be abolished too soon, at least not in his own lifetime.

Yet since gradualist modes of change structured Jefferson's thinking in many other contexts as well–even an "immediatist" document like the Declaration of Independence, after all, made a point of praising the patience of the Americans colonists[88]–it would be difficult to reconstruct the intention of dissimulation that would be necessary to explain his position on slavery by his hypocrisy. As will be discussed in the second part of this book, Jefferson was a scrupulous, history-conscious stylist who composed and preserved his writings with an eye to his future historical reputation. In his gradualist conception of progress, he thought that his posterity would largely move upwards along the gradations of scientific, political, and moral progress. It was essential to his understanding of the gradual enlightenment of mankind that he expected his future readers to be more enlightened than his own generation: he thought they would gradually learn more about nature and human nature, spread republicanism, and, sooner or later, abolish the institution of slavery. Thus, Jefferson must have expected later historians to judge him by standards that were morally and politically more progressive than those of his own society.

The fact that this posterity-haunted writer nevertheless failed so conspicuously to make plausible excuses for his views on slavery in the writings he preserved for future readers–at least attempting to mitigate what tends to strike latter-day readers as a disturbingly self-righteous tone in many of his statements on American slaves–suggests that he was not aware of the extent to which his gradualism would later be seen as morally compromised. He must have considered the position of his own generation enlightened enough to stand the test of time, more enlightened, at least, than the extremes of the unprogressive passivity of the prerevolutionary generation and the temporary regression of the "post-revolutionary youth." Therefore, character-centered speculations about Jefferson's personal bent toward hypocrisy appear, by themselves, too simple. Especially since gradualist arguments proved to lose credibility among the following generations, when men in his position would more often argue either for the maintenance or for the immediate abolition of slavery, Jefferson's gradualism also needs to be seen in the context of the problems of his late Enlightenment approach to change in time.

It would be unsatisfactory, to be sure, to make tautological claims that his thought and actions were entirely circumscribed by a particular concept of temporality. Yet if such claims can be avoided, the reconstruction of Jefferson's views on time and history helps us to better understand the Janus-faced character of a transitional worldview seeking to bridge the gaps that were being created by the secularization of Euro-American culture. As I have argued, Jefferson's gradualism was his answer to the question how to integrate discrepant experiences of time into his view of the (individual) good life. As part 2 will show, it also shaped his views on (collective) history and historiography, as it seemed to offer a viable compromise between different temporalities in his revolutionary "world of changes."

PART II

HISTORY

Jefferson's Newtonian temporalities provided the structure for a historical consciousness that is far from self-evident in today's more radically secularized world. As I shall argue in the following chapters, the tensions within his gradualism between a rational and a sentimental mode of time perception shaped his views of the past and his expectations of the future. A passage Jefferson read in *Tristram Shandy* discusses the connection between Locke's two temporalities and an enlightened concept of *history:*

> Pray, Sir, in all the reading which you have ever read, did you ever read such a book as *Locke's* Essay Concerning Human Understanding? . . . As I write to instruct, I will tell you in three words what the book is.–It is a history.–A history! of who? what? where? when? Don't hurry yourself.–It is a history-book, Sir, (which may possibly recommend it to the world) of what passes in a man's own mind; and if you will say so much of the book, and no more, believe me, you will cut no contemptible figure in a metaphysic circle.[1]

Tristram's "three-word"-definition of Locke's *Essay Concerning Human Understanding* as a "history-book" shrewdly revealed the ambivalent status of *history* in eighteenth-century thinking. It allowed for two different approaches, one relatively static, the other more dynamic. The first emphasized a rational temporality governed by natural laws; the second, a sentimental temporality based on human perceptions.

In the first context, it made sense for eighteenth-century minds to label Locke's *Essay* a "history," since the contemporary use of the term referred, not directly to past events or the past as such, but to a loosely defined genre of writing. This "historical" genre of writing, which Jefferson used as one of three categories in the Baconian classification of his library, consisted in descriptions of all known facts regardless of their position in time. It contained civil and natural history alike, including such (for modern sensibilities) largely atemporal and "unhistorical" subjects as surgery, chemistry, and agriculture.[2] The *Essay* could be called a "history" by Sterne and his contemporaries for its description of "natural" operations in the human mind taking place in an unspecific present that was largely similar to the past and future. In the second context, however, the definition of Locke's *Essay* as historical writing also anticipated a more dynamic concept of history. As a "history-book . . . of what passes in a man's own mind," the *Essay* differed from "natural history" in its connection to a specifically human subject matter, the internal passing of ideas in the mind. As discussed in the previous chapters, it was this internal "succession of ideas" that according to Locke was the basis for the perception of time. Hence, the identification of a "historical" subject matter with "what passes in a man's own mind" can also be read as a paraphrase for the subjective sense of the passing of time. If Locke's *Essay* was a "history-book," in this sense, then history was concerned less with the description of a universal nature than with the perception of temporal change. The definition of Locke's *Essay* as "history" in Jefferson's cherished *Tristram Shandy* therefore not only implied the traditional sense of a genre that also included natural history. In addition, it suggested an understanding of *history* as the description of events in a specifically human temporality, and possibly even of *history* as a term denoting these events themselves.

Eventually, aspects of the temporal dualism implied by Newton, Locke, and Sterne were united in a modern historical time. Its emergence was institutionalized, to some extent, at about the middle of Jefferson's life by the French decimal calendar that was in use from 1793 until 1805 but had been suggested as early as 1788.[3] Its most obvious political function was to break with the religious worldview of the *ancien régime* by replacing the Christian Era by a Republican Era and by substituting the *décade* for the seven-day week. Yet this revolutionary measure broke not only with the biblical week but also, and more importantly in Jefferson's context,[4] with the astrological week, suppressing the traditional names of planets and gods or goddesses by a more abstract decimal order. The names derived from celestial bodies were replaced by names derived from abstract numerals, which no longer referred to external movements in space but to operations within the human mind. Symbolically, time measurement was thus emancipated somewhat from its ancient connection to movement. The abstract naming of the weekdays emphasized a subjective dimension of the perception of time. This subjectivity, however, was now expressed in the universal order of mathematics, thus synthesizing aspects from the temporalities Newton and Locke had described. In this sense, the revolutionary calendar was a political equivalent to the new concept of time in transcendental idealism. Although less consistently than Kant's definition of time as a formal a priori condition of all appearances, the French innovation likewise associated the universal "mathematical time" of Newtonian philosophy with processes in the mind. Thus moving beyond the dualism of rational and sentimental time, it could refer to a modern time structured by mental operations but measured according to universal laws within an entirely secular human history.

In the American context, as is well known, such a calendrical reform never gained acceptance. And the flexible dualism of two times continued to shape Jefferson's political arguments. The resulting plural of different temporalities was condensed in the first seven words of the text with which he is most often associated, the Declaration of Independence: "When in the course of human events . . ." Instead of referring to this point in time by a more modern expression such as "when in the course of history," "when in history," or simply "when," Jefferson specified it by the curious eighteenth-century phrase "course of human events." By these words, he referred to a decisive historical moment in the summer of 1776 as part of something that was both singular and plural, at once exemplary and unique, an abstract whole as well as many concrete parts, a single, homogeneous, natural "course" that comprised plural, heterogeneous, unpredictable "human events."

The Jeffersonian combination of a natural "course" and "human events" contained the potential of two divergent conceptions of history. During his long life, Jefferson's eclectic thought oscillated between these two poles: On the one hand, there was the "course" pole, defined mostly in opposition to seventeenth-century antiquarian historians, a teleological concept of history that emphasized less the factual precision than the universal didactic function of historical studies. And on the other hand, there was the "event" pole, a concept that stressed the contingency of history and saw the primary function of historical studies not in teaching timeless moral lessons for the future but in coming as close as possible to a unique reality in the past. Theorizers of the "course" concept of eighteenth-century history tended to try to transcend the Aristotelian division between historiography and poetry by stressing the literary qualities of historical works and their aim to teach general rather than merely particular truths. They regarded historiography, as they did poetry, as an imitation of the probable that was primarily conveyed by the senses, rather than by reason. In the late eighteenth century, however, the aesthetic reading process of this mimetic history became more and more problematic. With the experience of an "acceleration" of time, the "event" pole began to move into focus, and a neo-Aristotelian history came to be seen as a problem with its own terms. History thus partook in the trend of fictional literature to become emancipated from the universal mimetic aim of the imitation of the probable, and each was enabled, in the long run, to assume a realm of its own.

Until now, Jefferson's concept of history has been examined mainly from the point of view of political theory as well as political and constitutional history. His historical thought is thus mostly known for its republican content–as gained from his reading of the Whig historians[5]–and its immediate functions in his rhetoric. Due to this emphasis on the direct political import of his historical arguments, other aspects of his historical thinking have not been studied in detail. If Jefferson's attitude toward theoretical and methodological questions of history has been described, it has generally been assumed that he held "typical" Enlightenment views–approaching history in neoclassical fashion as a collection of timeless examples for the practical use of future generations, failing to differentiate between the "experience" of the past and the "experience" of the present, and assuming transhistorical laws to be at work behind civil and natural history alike. In other words, if Jefferson's historical arguments have been analyzed, the tendency has been to associate them with the "course" pole of his thought.

However, Jefferson's writings were complicated by a qualifying temporality, at the "event" pole of his historical thinking. In a period that was collectively experienced as a time of radical political and cultural upheaval, a didactic historiography relying on classical examples that merely quantified change did not always offer satisfactory modes of explanation. As a consequence, Jefferson's statements on history were characterized by striking ambiguities, offering at least two different solutions to many important theoretical and methodological problems. For instance, he was uncertain about the role of literary style in works of history. While he cherished stylistic brilliance in ancient and modern historiography throughout his life, he also came to fear the seductive effects of the "magic" of the literary style of historical works. Jefferson's statements were equally ambivalent about the role of factual accuracy. He sometimes implied that historical details could be neglected if they were unable to teach a larger truth, while in other contexts he explicitly complained about their neglect and considered them interesting in their own right. In like manner, he sometimes advocated the use of classical costume for the visual representation of historical characters in painting and sculpture, while on other occasions he found everything ridiculous but a period costume, conveying the accidental details of a particular historical situation. His conception of historical characters in writing was similarly undecided. Jefferson sometimes understood them primarily as moral examples (almost as natural types), personifying universal truths about political reality, while at other times he characterized them as "singular" individuals. Moreover, he made contradictory statements about the projected audience of historical works. While he usually implied the classical notion that history was written for a timeless posterity, he sometimes showed himself aware that historical works were really written for a contemporary audience, whose needs and interests would change over time. The ultimate uncertainty may be found in Jefferson's reflections on the intellectual status of historical studies. He often assumed that they involved the senses and "principally the memory" and could therefore be regarded as "lighter occupation" to fill the time after dinner. In this train of thought, he imagined readers to be passive consumers who could be "nourished" or "poisoned" by historical works. Especially in his later years, however, Jefferson also began to imagine historical studies as a more active process, claiming that they involved not only the memory and the senses but also rational "research and reflection" on the part of both readers and writers.

These conflicting assumptions about history existed side by side in Jefferson's intellectual repertory and could be emphasized in various combinations depending on the occasion. In a mind as self-consciously eclectic as Jefferson's, of course, it is not surprising to find what strike us today as contradictory conceptions coexisting peacefully in one and the same text (as in his *Notes on the State of Virginia*). However, his different approaches to history also tended to have different degrees of relevance in different situations of his life. Instead of treating Jefferson's thought as an essentially static whole (which is the more frequent practice), the following chapters are therefore structured chronologically. With a conception of historical change as their subject matter, they seek not to avoid, themselves, the challenge of reflecting on the role that the great upheavals of the revolutionary period played in Jefferson's thought. Simplifying matters somewhat, the three chapters accordingly describe three different emphases in Jefferson's historical thought as they converged, more or less, with three major phases in his life. I begin with the young Jefferson's historical frame of mind and his relatively uncomplicated expectation that history was a collection of timeless examples following a "course" predictable by universal natural laws. His experience of chaotic "human events" in the revolutionary period, however, posed a serious threat to the predictive function of history. Especially in the late eighteenth century, Jefferson tended to express a greater skepticism about the Thucydidean aim of political foresight, beginning to see in past experience, to some degree, a value and a standard of its own. This potential conflict between different expectations of history was partly resolved in Jefferson's thought by the event he identified as the "Revolution of 1800." As he wrote in 1801, after his election to the presidency, everything in America was "new." It was this American "newness" that made it possible for Jefferson both to transform and to consolidate his early assumptions of a didactic American historiography.

Chapter 4

Teaching by Examples

The historical thinking of the young Jefferson was shaped by the philosophical history of the Enlightenment,[1] a conception perhaps best summarized by an aphorism of Henry St. John, Viscount Bolingbroke, defining history as "philosophy teaching by examples." In a book that was part of Jefferson's library, *Letters on the Study and Use of History,* published in 1752, Bolingbroke recommended the study of history as "the most proper to train us up to private and public virtue."[2] Contrary to the supposedly useless pedantry of seventeenth-century antiquarian historians, writers like Bolingbroke tended to see the primary aim of history less in preserving the details of a particular historical reality than in teaching universally valid moral lessons about a transhistorical concept of "human nature." Referring back to axioms developed in antiquity, Enlightenment thinkers tended to conceive of history as a collection of exemplary narratives about the past meant to be of immediate use for the moral and political instruction of present and future generations. This timeless didactic function makes philosophical history appear today–as the somewhat self-contradictory label might express[3]–as a particularly "unhistorical" form of historical study. Its declared aim of discovering "the constant and universal principles of human nature," to quote David Hume,[4] lost plausibility with the rise of "Romantic" historiography, with its contrary emphasis on the individual and contingent aspects of human experience. It is therefore necessary to reconstruct precisely how Jefferson and his contemporaries were able to endow history with a universal moral utility.

As Jefferson never developed a systematic "theory" or "philosophy" of history himself, his statements on the past and on historical studies have to be analyzed in context with several other authors. On the one hand, there were figures like Bolingbroke and Henry Home, Lord Kames, who were relevant to Jefferson's historical thinking, as they popularized typical assumptions of philosophical history that were immediately important in his formative years.[5] On the other hand, there were philosophers whose more abstract reflections may not have concerned the young Jefferson directly but who discussed crucial problems in enlightened approaches to the past that would also determine the framework of his arguments. The most important author in this context was David Hume, whose writings contain one of the most complex theories of philosophical history, pointing toward tensions in the empirical Enlightenment that ultimately destabilized Jefferson's historical thinking as well. Although Hume did not always follow his own theories in his historiographical practice,[6] his *History of England* (1754–62) would, moreover, become one of the touchstones for Jefferson's later relationship to Anglo-American history.[7]

History in a Clockwork Universe

The didactic aim of philosophical history could best be achieved within the order of a clockwork universe, based on an idealized conception of nature in rational time. As has been said, this transitional temporality was no longer the inferior, "merely" human time of traditional Christianity, and it was not yet the secularized historical time of the nineteenth and twentieth centuries. Rational time was supposed to be steadily progressing from the past into the future according to universal laws of nature that were assumed to be independent of human actions and perceptions. From today's perspective, this beginning secularization of eternity and the resulting absolute, almost superhuman character of rational time accounts for the impression of a peculiar timelessness in eighteenth-century reflections on history. What appears to modern eyes as timeless, however, did not result from a failure to reflect on time in general, but only from a lack of reflection on a uniquely human temporality. As enlightened conceptions of time presupposed a purposeful universal order, all kinds of nature–mechanical, organic, and human–could be regarded as subjected to the same absolute, "equably" flowing time that Newton had described. Such a homogeneous temporality could allow for a certain degree of quantitative change–the gradual fulfillment of a "design" in nature, for instance[8]–but had difficulty integrating radical changes occasioned by previous events that were different in quality from those of the present.

In this temporality, the philosophical historians did not usually draw a clear line between human motives and actions, on the one hand, and what they often called "the course of nature," on the other. On the contrary, they tended to stress that the same laws were at work behind civil and natural history alike. In their enlightened "world of laws, measurements, predictions, and constancies or regularities of behavior," as Gordon Wood has described it,[9] civil history could be imagined, in principle, as forming regular patterns comparable to those discovered in natural philosophy. Associating "human events" with natural events like the "revolutions" of planets in astronomy,[10] the philosophical historians tended to expect the future to follow largely the lines prescribed by the past.[11] Like planets moving in their regular orbits, human events could thus be expected to form a discernable "course of human events" and even an "order of events," to use Jefferson's characteristic paraphrases for "history" in the Declaration of Independence and in *Notes on the State of Virginia.*[12]

In a clockwork universe, human nature had as little room for sudden and arbitrary change as the rest of nature. Accident or chance and unforeseen rapidity–sometimes illustrated by the astronomical metaphor of a "meteor," as in Jefferson's and Bolingbroke's writings[13]–were not completely absent from this worldview, but played a relatively marginal role in its historical explanations.[14] If events appeared to deviate randomly from the supposed regularity of nature, the philosophical historians did not tend to make great efforts to explain them on their own terms. Instead, they often tried to account for them as being caused by secret "springs and principles" that did not contradict the universal rule but were merely "hidden"–like the invisible Saturday of Jefferson's Great Clock–from the historian's limited human perspective. Typically, "irregular events" were thus explained less by a "contingency in the cause" than by "a secret operation of contrary causes," as Hume argued using the example of a defective clockwork.[15]

In its aim of discerning universal patterns behind particular incidents of the past, philosophical history presupposed a relatively rigid notion of causality. Whether or not this causality was identified, following the Humean argument, as connecting principle in the mind (and in writing) instead of the events themselves, it was often represented by a spatial image condensing this rigidity: the hierarchical "great chain of being" thought to compose the subject matter of natural history was paralleled by the "great chain of events, which compose the history of mankind," as Hume put it.[16] In a book that Jefferson commonplaced, Hume's relative Lord Kames spoke of "a regular chain of causes and effects" and described how historians, in order to reconstruct this "regular historical chain," had to "supply the broken links, by hints from poets and historians, by collateral facts and by cautious conjectures."[17]

This conjectural method of gaining historical knowledge by supplying "broken links" in the "regular chain of causes and effects" was tied to the assumption of a universal human nature. A search for links in the "chain" of history was grounded in the "universally acknowledged" assumption, as Hume wrote,

> that there is a great uniformity among the actions of men, in all nations and ages, and that human nature remains still the same, in its principles and operations. The same motives always produce the same actions: The same events follow from the same causes. Ambition, avarice, self-love, vanity, friendship, generosity, public spirit; these passions, mixed in various degrees, and distributed through society, have been, from the beginning of the world, and still are, the source of all actions and enterprises, which have ever been observed among mankind. Would you know the sentiments, inclinations, and course of life of the GREEKS and ROMANS? Study well the temper and actions of the FRENCH and ENGLISH.[18]

The metaphor of history as a homogeneous "chain of causes and effects" would have been implausible without the supposition of a general human "uniformity" remaining at bottom unchanged by the passing of rational time. Only then could "the same events" in the future be expected to "follow from the same causes" as in the past. Human actions were thought to reveal a number of essentially unchanging passions, invariable "principles" of a uniform human nature that had remained the same since "the beginning of the world." This uniformity, however, did not mean a "uniformity in every particular," which Hume thought–and Jefferson would follow him in this–could be "found in no part of nature."[19] The philosophical historians universalized physical and moral aspects of human behavior but also allowed for a certain amount of variation in time and space. Based on this flexible universalism, they tended to see themselves as students of a repertory of transhistorical human attributes exposed to what they often called different "circumstances." Essentially "uniform" characteristics were subjected to environmental factors responsible for a particular historical distribution "in various degrees . . . through society." As the differences between "nations and ages" were considered to be gradual rather than essential, historical events could be understood as timelessly valid moral examples, instructing past, present, and future generations in the essentially uniform attributes of human nature. Knowledge of present "human events" could be used to explain the past, and knowledge of the past could be directly applied to one's own behavior.

This exemplary function of history typically found expression in static and spatial, rather than dynamic and temporal, metaphors. Apart from the rigid "chain," history could also be illustrated by images from painting and especially, in the tradition of ancient mnemotechnics, architecture. Among the most intriguing contemporary metaphors for history was that of a boudoir of mirrors:

> The History of Greece should be to our countrymen what is called in many families on the Continent, a boudoir, an octagonal apartment in a house, with a full-length mirror on every side, and another in the ceiling. The use of it is, when any of the young ladies, or young gentlemen if you will, are at any time a little out of humour, they may retire to a place where, in whatever direction they turn their eyes, they see their own faces and figures multiplied without end. By thus beholding their own beautiful persons, and seeing, at the same time, the deformity brought upon them by their anger, they may recover their tempers and their charms altogether.[20]

In language reminiscent of Adam Smith's description of the social mirror,[21] John Adams thus illustrated the relevance of ancient Greek history to Americans in 1787 by the complex image of a boudoir in the shape of an octagon, a shape symbolizing eternity that Jefferson appreciated in his own architectural designs.[22] Adams referred to history not as an abstract process in time but as a highly visual (though partly illusory) space. In many ways typical of eighteenth-century historical thought, the "History" of this description was not a temporal sequence of different events but an extension of almost identical figures in space. Although "multiplied without end," this spatial chain of reflections consisted of figures that were visible "at the same time" to the historical spectator, who could, moreover, enter the octagonal boudoir "at any time." The simultaneity, infinity, and near identity of the different visual impressions he or she received in the endless reflections of the mirrors were based on the quasi-eternal homogeneity of rational time, which allowed for a certain degree of quantitative change (the size of the individual reflection would seem to gradually diminish in space) but tended to preclude qualitative change (the mirrors always reflected the same reflections of the same passions reflected in the same face). The rather claustrophobic character of the spectators' historical experience of seeing their own "deformity" in endless historical successions, wherever they turned, had the didactic function of improving their behavior in the future. As indicated by the mirror metaphor in connection with a "young lady," it was ultimately an allegorical allusion to *vanitas*–the spectator's vanity combined with the consciousness of her own finite nature as a mere link in the chain of historical reflections–that was supposed to achieve the beneficial effect of self-control.

This humorous passage left open whether moral progress could result from such a self-centered, self-reflective historical education. While Adams, in any case, understood Greek history to present merely "anger" and "deformities," his young spectator could only be "beautiful" if she was able eventually to "recover" her temper. In similarly ambiguous ways, Bolingbroke had described human vanity as the main motive for studying history. As he put it in his *Letters,* "The love of history seems inseparable from human nature, because it seems inseparable from self-love. The same principle in this instance carries us forward and backward, to future and to past ages."[23] In the timeless historical space of the "boudoir" of philosophical history, the spectator's position was simultaneously backward- and forward-oriented: the young lady's "retirement" to her historical reflections was aimed at the "recovery" of an original state in the future.

Thus, Adams's image of a mirrored boudoir illustrates at least four important problems of philosophical history: first, the spatial extension rather than temporal succession of a characteristic eighteenth-century way of imagining the past; second, a beginning consciousness of the mediated, potentially illusory character of history understood as a series of visual impressions made by mirror images judged by their aesthetic qualities; third, the tautology of universalist assumptions about a past human nature imagined as mere reflection of one's own behavior; and fourth, a paradoxical simultaneity of a past- and future-orientation, implying the conception of the future as a "recovery" of an original state in the past. In what today appears rather a circular manner due to the tautology of a uniform human nature, the comparative methods in philosophical history of observing universal "principles" of human behavior in their different historical gradations made the philosophical historians think in both directions of rational time and of the "regular chain of causes and effects." They could use their knowledge of human behavior to make "cautious conjectures" from known effects to unknown causes in the past. Thus improving their understanding of human nature further, they were able by analogy to draw conclusions from known causes in the past to unknown effects in the future. Their increased causal knowledge of past events could therefore eventually be used as exemplary knowledge about supposedly similar events in the time to come. Hume formulated the didactic essence of the "regular chain" idea of history by characterizing the narration of the causal links in the "great chain of events" in historical works as "the most instructive, since it is by this knowledge alone, that we are enabled to controul events, and govern futurity."[24]

As this terminology indicates, rational time in philosophical history retained its implications of social and political control. The aim was to "govern futurity," routinely using knowledge of the past to predict the future and learning "lessons" from history in order to better "controul" future politics. The chief use of philosophical history was less to acquire retrospective knowledge about a particular quality of the past than to increase the quantity of historical information in both directions of rational time. Looking toward the past, the philosophical historians had the paradoxical aim of improving their perspective of the future and forming the socially and politically relevant habit of foresight that Jefferson, as has been discussed, also tried to impart to his family on the personal level.

Foresight, Continuity, and Restoration

The future-orientation of philosophical history, its supposition of transhistorical human characteristics, and its political function can be interpreted as continuing the classical Thucydidean vision of history.[25] Rooted in this tradition, the aim of political foresight was crucial to Jefferson's historical thinking. It shaped his understanding of the past as a young man and, after being complicated by fin-de-siècle periods of skepticism, again became important in his presidential and retirement writings, although on a new and more complex level. Written before his doubts about a historically grounded foresight intensified in the last years of the eighteenth century, Jefferson's public and private papers from the 1770s and early 1780s contain expressions of what may have been his least complicated reliance on the axioms of philosophical history: the universality of human nature, the resulting predictability of historical causes and effects, and the conception of the future as the return to an idealized past. For instance, in a letter to his former teacher William Small in May 1775 he claimed that the outbreak of violence at Lexington and Concord could have been "foreseen" and prevented had George III only had a "little knowledge of human nature" and paid "attention to its ordinary workings."[26] Based on the same concept of political foresight, Jefferson argued under query 13 of *Notes on the State of Virginia* that his fellow Virginians, in order to guard themselves against corruption and tyranny, had to "look forward to a time, and that not a distant one, when corruption in this, as in the country from which we derive our origin, will have seized the heads of government."[27]

Echoing Hume and Kames, he explained the predictability of political corruption in terms of the "regular chain of causes and effects": "Human nature is the same on every side of the Atlantic, and will be alike influenced by the same causes. The time to guard against corruption and tyranny, is *before* they shall have gotten hold of us."[28] Although he would qualify the sameness of human nature under the same and the following queries,[29] Jefferson's conviction of universal invariables in the study of mankind–its "inherent and inalienable rights" and "common feelings"[30]–was a crucial part of his worldview. He saw human nature as subjected to the universality of rational time on two levels, the prescriptive level of value and the descriptive level of facts. Like many of his contemporaries, he was usually not very precise in separating the two levels.[31] Thus, whether interpreted mainly in the relatively more prescriptive understanding of human nature in the "liberal" tradition of political thought or in the more descriptive understanding of human nature in the "republican" tradition, Jefferson's approach to the historicity of human behavior was defined by a number of universalist assumptions. If he understood human nature prescriptively, there were the "inherent and inalienable rights of man," which he tended to characterize collectively as "unchangeable" (in contrast to the historical contingency of legal rights).[32] And even if he understood human nature primarily as a descriptive category, it was still defined by certain transhistorical invariables that, although dynamic in themselves, were caused by similar factors in all periods. For instance, even in his classical "republican" utterances, he thought that the universal danger of political "corruption" was the effect of human weaknesses that were largely identical in all periods: an essentially timeless ambition, avarice, venality, desire for luxury, and so on.

To some extent, the inner dynamics of these classical republican vices were instrumental in preparing the ground for a modern historical thinking.[33] Yet Jefferson's emphasis on these republican vices as such was still part of a premodern conception of history, as he assumed that they existed equally in all modes of rational time. Remaining basically the same throughout history, like the mirror images in Adams's historical "boudoir," they pervaded both the past and the present of Jefferson's world, often threatening to affect the future as well. Their historical continuity made it plausible for him to allude to them in his own political experience in a neoclassical manner by historical "examples," references to supposedly similar cases in the past employed for the rhetorical use of illustrating a certain universal moral content.[34] Two instances of Jefferson's early reliance on the timeless exemplarity of history can be found under query 13 of *Notes on Virginia:*

> 173 despots would surely be as oppressive as one. Let those who doubt turn their eyes on the republic of Venice. As little will it avail us that they are chosen by ourselves. . . . Mankind soon learn to make interested uses of every right and power which they possess, or may assume. The public money and public liberty, intended to have been posited with three branches of magistracy, but found inadvertently to be in the hands of one only, will soon be discovered to be sources of wealth and dominion to those who hold them. . . . With money we will get men, said Caesar, and with men we will get money.[35]

Jefferson illustrated his criticism of what he called a possible "elective despotism" under the Virginia Constitution of 1776 by the quasi-visual example of the Venetian Republic, trying to persuade doubters of his statement by the immediate information of their "eyes." For him, the republic of Venice and republican Virginia were equally subject to the universal rule that without sufficient checks and balances between the separate branches of government, free elections could not prevent "despotism." Once this truth was made visible rather than being "inadvertently" overlooked, Jefferson assumed, political corruption might be foreseen and perhaps prevented, as he expected it always to take essentially the same course. The structure of his universalist argument about the dangers of money was similar: based on his knowledge of "Mankind," he explained the permanent factors of human "interest" and "venality" in the course of political corruption by the historical example of Caesar, for Jefferson the timeless personification of the turn from republicanism to "despotism." Despite the location of these historical examples in the ancient past, significantly, much of this part of query 13 is in the future tense, emphasizing the Thucydidean goal of political foresight in philosophical history.

In this continuous understanding of the past, Jefferson tended to naturalize historical categories. A historical agent like Caesar functioned almost as a natural type, his classical costume conveying less the details of his period than one of the "constant and universal principles of human nature" at work in Jefferson's own experience. A reference to the Renaissance or to antiquity was thus able to signify that something was still present or even future, rather than irretrievably past. The continuity of history was least complicated in Jefferson's aesthetic preferences, his neoclassical admiration for antiquity in architecture and writing. As in the opposition between the early Georgian architecture that Jefferson sometimes derided as "Gothic" and the neoclassical architecture that he admired and practiced,[36] classical references conveyed less a backward-oriented historicity than a timeless naturalness and even a forward-oriented modernity to eighteenth-century minds. Jefferson's tendency to recommend buildings from the Roman Empire as models for American buildings demonstrates that he could not have been interested, in this context, in the particular political quality of a past that he usually placed among the most morally corrupt in history.[37] When he used such buildings from the empire as the Pantheon or the Maison Carrée as models for modern American architecture, he did so not because of their historical particularity but because of their universal aesthetic quality. Jefferson grounded his universalist aesthetics in the approbation "by the general suffrage of the world" of a building from the Roman Empire that had "pleased, universally, for near two thousand years."[38] Although not as pronounced as in art history, natural-historical invariables like the exemplary despot Caesar contributed to the naturalization of political categories. Jefferson's natural metaphors for political life, whether in his images of wolves and sheep for political oppression[39] or in his claims that "in fact the terms of whig and tory belong to natural, as well as to civil history,"[40] similarly partook in a continuous classical world of transhistorical causes and effects.

As I argue in the following chapters, Jefferson's conception of history lost some of its pronounced universalism in his middle years and regained it only with qualifications in his old age. His understanding of historical continuity was challenged, ultimately, by his evolving thoughts on moral and political progress. While his reading about the progress of civilization tended to undermine the idea of a stationary human nature from early on, it was the radical "newness" of his historical experience of the last decades of the eighteenth century and especially of the "Revolution of 1800" that contributed most to the modification of his thought. This modification may be briefly illustrated here by the evolution of his Whiggish relationship to Anglo-Saxon England. After he recommended the Anglo-Saxon chiefs Hengist and Horsa for the reverse side of the Great Seal of the United States in 1776 and spoke of Anglo-Saxon England as "that happy system of our ancestors, the wisest & most perfect ever yet devised by the wit of man, as it stood before the 8th century," he doubted by 1799 "that nothing can ever be devised more perfect than what was established by our forefathers," whom he finally characterized as "a semi-barbarous ancestry" in 1824.[41] Although his admiration for the Anglo-Saxons continued throughout his life, his ability to historicize them apparently increased, while his willingness to regard them as part of a historical continuum, as timeless examples directly applicable to the present situation, would be considerably diminished in the nineteenth century.

It was thus in his early years that Jefferson made his strongest statements of a historical continuum between past, present, and future in rational time. Emphasizing the uniformity rather than the variations of human nature, the young Jefferson could even imagine the future as a "return" to or a "restoration" of the past. A year before he suggested a "return at once into that happy system of our ancestors," he had expressed hope for "a restoration of our just rights" as well as for "a return of the happy period, when, consistently with duty, I may withdraw myself totally from the public stage, and pass the rest of my days in domestic ease and tranquillity, banishing every desire of ever hearing what passes in the world."[42] Thus, the younger Jefferson was able to envision the future as the repetition of an idealized past on both the political and the personal level–of the quasi-natural state of Anglo-Saxon liberty before the Norman Conquest and the natural idyll of a "tranquil" domestic happiness before the historical "confusion" of the American Revolution.[43]

As in the writings of many contemporaries, to be sure, this call for a "return" to a mythical golden age before the intrusion of "history" is not to be taken at face value. It served different rhetorical needs and is seldom to be read literally–not even in the writings of Rousseau and certainly not in those of Jefferson, for whom the actual "reduction" to a "state of nature" clearly had negative connotations. It is remarkable, however, that the early Jefferson could cling to this theme of "return" and "restoration" quite stubbornly if it supported his argument. A case in point is a complex expression of his early historical thought, his *Summary View of the Rights of British America.*[44] In this text, which can be characterized as "transitional" in its shifting conception of rights[45] as well as of history, Jefferson's reasoning rested on the natural homogeneity of past, present, and future, even while stressing the crisis in the present.[46] In his description of this crisis, he began to move toward the sense of historical uniqueness sometimes felt in his later writings, while also taking care not to stretch philosophical history to its limits. When Jefferson described recent events as characterized by suddenness ("the whole of that antient and wealthy town is *in a moment* reduced from opulence to beggary") and unprecedented rapidity ("that rapid and bold succession of injuries which is likely to distinguish the present from *all* other periods of American story"), the regularity of rational time was beginning to become problematic. Yet despite their "irregular" acceleration, these events did not in the end create "new" historical conditions or modes of action, but only amounted to "extraordinary situations which require extraordinary interposition"–with the implication, of course, that one could indeed still expect other historical situations or interpositions to be "ordinary."[47]

Thus, although the natural order of history was beginning to become doubtful, the *Summary View* still emphasized a basic "nature of things" that had stayed and would stay the same "at all times." Jefferson's interpretation of rights gained through expatriation, by a transhistorical "universal law" affecting the emigration of Anglo-Saxons and British colonists "in like manner," made him emphasize not only the continuity but even the near identity of the two cases, separated as they were by more than a millennium: "And it is thought that no circumstance has occurred to distinguish materially the British from the Saxon emigration." In this sentence, Jefferson blended fact and value, which were beginning to diverge in other parts of the text. Not only had the "universal laws" remained the same concerning the Anglo-Saxon and the British emigration but even the material "circumstances" of the two cases had supposedly remained identical. In such a static universe, homogeneous on the prescriptive as well as on the descriptive level of human nature, it was only natural to "sacrifice every thing which reason can ask to the *restoration* of that tranquillity for which all must wish."[48]

Considering the long list of historical violations of rights Jefferson presented in the *Summary View*, the ideal period of "tranquillity" he could hope to see "restored" in the future appeared to be located at quite a remote time in the past, although the text is ambiguous as to its precise historical location. It was in any case useful for Jefferson to go back in his argument, in Whiggish manner, to Anglo-Saxon England, as its remoteness in time and shortage of historical documents made it a smooth projection surface for a possible content of "restoration," his professed desiderata of "tranquillity" as well as "fraternal love and harmony." Its historical vagueness made the myth of Anglo-Saxon liberty verge, in a sense, on the idea of a state of nature. The allusion to such a quasi-natural state in a historical argument allowed Jefferson to establish the identity of fact and value in the past that was characteristic of the naturalizing tendencies in philosophical history.

The future perspective of Jefferson's "restoration" was symmetrical to this half-historical, half-natural past: as made explicit by his emphatic claim in the last sentence of the *Summary View,* it was supposedly "the fervent prayer of all British America" for this ideal state, once restored, to "continue to the latest ages of time." Thus, his early view of Anglo-American history was characterized, not by the expectation of fundamental change, difference, and "newness," but by an emphasis on continuity and identity at both ends of the straight line of rational time, an imagined stability in the remote past and the remote future. This suggests that Jefferson did not argue exclusively either "from history" (i.e., from the contingency of historical precedent) or from a universalist natural-rights philosophy in the *Summary View.* Instead, he combined the two approaches: making some allowances for historical contingency in the crisis of the present and the more recent past, he envisioned a long-term future that was defined by a "nature of things" that had already existed ages ago.

Despite the provocative function of the text as an ultimatum to the king for a change of policy, the last sentence of the *Summary View* underlined the long-term stability of Jefferson's early conception of a continuous British and American history, which, as he saw it, was unchanged in its essential "principles" by the passing of rational time.[49] In this historical interpretation, of course, it was the king, and especially his "British councellors," rather than "British America," who had rebelled against the universal "nature of things." In order to achieve a "restoration," it was accordingly the king who had to learn and improve his behavior. The *Summary View* addressed him less as the customary father figure–Blackstone's *paterfamilias* of the nation[50]–than as a prospective student of philosophical history. An expression of this irreverence toward the monarch, the text presented a detailed statement of how Jefferson imagined this historical learning process to take place. It can be read as an imagined lesson in this learning process, a history course of "philosophy teaching by examples" with George III as its sole student.[51] This student, tutored by a paternalizing "great American council," was supposed to improve his proficiency in philosophical history because of its two major functions: to teach exemplary lessons and to pass moral judgments. The *Summary View* admonished George III not to look backwards and "imitate" the wrong examples of his predecessors but rather to look forward to his own short- as well as long-term future and consider that his "felicity and future fame" might depend on his willingness to learn the crucial historical lessons that the text presented.[52] His future place in history should be the motivation to "expand" his historical knowledge of human nature: "Let those flatter who fear; it is not an American art. To give praise which is not due might be well from the venal, but would ill beseem those who are asserting the rights of human nature. They know, and will therefore say, that kings are the servants, not the proprietors of people. Open your breast, sire, to liberal and expanded thought. Let not the name of George the third be a blot in the page of history."[53]

This famous passage, set off stylistically from the rest of the *Summary View* by its relatively short sentences, can also be read as a statement that the king's "great American council" was distinguished from his "British counsellors" mainly by being composed of the better historians or teachers of philosophical history. It was composed of independent people who were not subjected to Montesquieu's tyrannical principle of "fear" and did not give flattering historical "praise which is not due," but said what they knew about a universal human nature and accordingly passed the just historical judgments by a universal standard, whose supposed impartiality was underlined by Jefferson's generous advice to the king: "Only aim to do your duty, and mankind will give you credit where you fail."[54] Thus, as was typical of philosophical history, the "expanded thought" that the lessons of the *Summary View* offered to George III expanded in both directions of rational time–to the remote past going back to the Anglo-Saxon emigration and to the remote future of the king's supposedly permanent character in historical writing, in the (single) "page of history" that would be dedicated to him. Based on his new insights into the continuity of British and American history that the *Summary View* provided, George III was supposed to gain political foresight and contribute to the "restoration" claimed necessary by his American subjects. In Jefferson's following writings, this procedure was republicanized: instead of imagining the king as a history student, Jefferson began to think about history as a key subject in the education of Virginia citizens.

A Moral Exercise

In Jefferson's early concept of history, which envisioned the remote future, to some extent, as the reenactment or "restoration" of a remote past, it was still a relatively uncomplicated assumption that the knowledge of times past could be directly applied to improving the times to come. To achieve this end, a historical narrative like the *Summary View of the Rights of British America* could provide moral examples, pass moral judgments, and allude to future moral judgments. At least for the sake of rhetoric, the writer of the *Summary View* thus expected his royal reader to learn from the past for the future. Yet Jefferson's professed optimism in the instructive power of history soon received its first blow. As he made clear two years later in his draft of the Declaration of Independence, his history student had utterly failed to learn from past examples and establish his "future fame." Instead, as far as Jefferson was concerned, George III had forever ensured his status as a "blot" in the historical record. He had done so not by an accidental error but–a point that is even stronger in Jefferson's draft than in the Congress version–by a homogeneous, unbroken chain of intentional measures designed to establish "absolute tyranny" in America.[55] In the universalizing understanding of historical causality characteristic of philosophical history, Jefferson considered the king's behavior to be the direct result of his motivation, and his motivation–as the *Summary View* had emphasized–to be the result of his historical education. The events leading up to American independence thus appeared to have been ultimately caused by the monarch's stubborn resistance to the correct interpretation of the past.

After his efforts to educate George III proved hopeless, Jefferson turned his attention to republican citizens in America as students of philosophical history. During the Revolutionary War, he promoted history as a key subject of public education in Virginia, mainly in two texts: the *Bill for the More General Diffusion of Knowledge,* which he drafted in 1778,[56] and *Notes on the State of Virginia,* which he began to compose about 1780.[57] In contrast to the conventional concept of history as a proper subject of study for gentlemen, noblemen, and statesmen–as, for example, Locke, Bolingbroke, and Lord Chesterfield had described it–Jefferson now began to think of it as the most important subject for the basic education of, significantly, "the great mass of the people."[58] Under section 1 of the bill, he explained this preference for history as a basis for educating "the great mass" in the terms of philosophical history:

> Experience hath shewn, that even under the best forms, those entrusted with power have, in time, and by slow operations, perverted it into tyranny; and it is believed that the most effectual means of preventing this would be, to illuminate, as far as practicable, the minds of the people at large, and more especially to give them knowledge of those facts, which history exhibiteth, that, possessed thereby of the experience of other ages and countries, they may be enabled to know ambition under all its shapes, and prompt to exert their natural powers to defeat its purposes.[59]

A historical lesson that Jefferson had learned by 1778, one could say, was that if an individual student of philosophical history had failed, it was all the more necessary to educate "the great mass of the people" in this field to compensate for such a failure. Political foresight should not be limited to "the rulers of the people," who had proved prone to corruption also in the recent past, but should be acquired by "the people themselves." Anticipating what would become a typical trait of his thought, the *Bill for the More General Diffusion of Knowledge* and the *Notes on Virginia* recommended a broad public education as an "essentially necessary" means to counter threats of republican corruption and tyranny.[60] In his opinion, education of "the great mass" could best contribute to preventing or at least temporizing the "degeneration" of government by rendering the people the "safe . . . guardians of their own liberty."[61] As Jefferson went on in the *Notes on Virginia,* "And to render even them safe their minds must be improved to a certain degree." This moderate mental improvement "to a certain degree" (and "as far as practicable") was supposed to be "chiefly historical." It consisted in the universal moral judgments and lessons of philosophical history: "History by apprising them of the past will enable them to judge of the future; it will avail them of the experience of other times and other nations; it will qualify them as judges of the actions and designs of men; it will enable them to know ambition under every disguise it may assume; and knowing it, to defeat its views."[62]

The popular knowledge of political "experience" in the past was supposed to help impede the repetition of past mistakes and make political progress possible. Jefferson thus laid "the principal foundations of future order," as he said, in the "first stage" of public education. This education, to be sure, was supposed to provide for the selection of his natural aristocracy. For the vast majority of the children included in his proposals, however, it would be limited to three years of gratis school attendance. In these years the boys and girls would have to learn reading, writing, and arithmetic, as well as what Jefferson modestly referred to as "the most useful facts of Grecian, Roman, European and American history."[63] Considering that he also expected this public education to qualify "the great mass" as "judges" of human behavior, making use of their newly gained historical knowledge in present politics and even future political decisions and actions, the primary stage of this instruction would have been rather intense for the students, to say the least. From today's perspective, such high expectations for a public three-year pre-grammar-school education appear almost ridiculous. The enormous benefits of the intended result contrast strangely with Jefferson's belief that his means to achieve this end were rather moderate and realistic, illuminating the public mind only "to a certain degree" and "as far as practicable." How could he seriously expect that three childhood years of rudimentary training in math and history for "the great mass of the people" would "qualify them as judges of the actions and designs of men" and become the crucial factor for the stability of the republic?

Jefferson's line of thought rested on a peculiar understanding of *experience,* which was often synonymous with, though not consistently, *history* as it is used today.[64] A flexible term, *experience* was able to encompass references to the knowledge of events in the remote and recent past as well as in the present, if they were seen as part of an all-inclusive empirical reality in rational time. This reality was not as complex as might be expected. In philosophical history, it varied over time in degree but not in kind. For instance, Jefferson thought that it was the goal of historical studies to "know ambition under all its shapes and defeat . . . its purposes" or "to know ambition under every disguise it may assume; and knowing it, to defeat its views." The assumption was that ambition had always been in essence the same, although "in other times and other nations" it had appeared in different costumes or even, according to the paranoid style of the late eighteenth century,[65] in different "disguises" (a formulation in which its various historical shapes almost amounted to intentional tricks devised by the "cunning" and "wickedness" supposed to accompany "ambition"). As a transhistorical link in a homogeneous "chain of causes and effects," it remained identical in its destructive effects for republican virtue. Jefferson's plans for public education thus did not imply that the "great mass of the people" should study the intricate differences between human motivations in history. It sufficed for them to learn what stayed the same over time in order to be able to react to it in their own political experience.

Like many of his contemporaries, Jefferson considered this learning process of identifying unchanging historical patterns in a continuous empirical reality to be important politically but not very demanding intellectually. Whether gained from the study of past events or from personal participation in the events themselves, experience was transmitted by sense impressions and, following Bacon's division of knowledge, memory, not necessarily by reason. "Chiefly historical" studies involving mainly sense impressions and memory could therefore be considered especially well suited for young children. The Jefferson of the *Notes on Virginia* felt that their memories should be cumulatively "stored" with useful historical facts while their minds were still relatively vacant. He found it irresponsible, by contrast, to make children read the Bible, since in his opinion religion, unlike history, required mature rational judgment.[66] In the spirit of Locke's and Bolingbroke's educational writings, Jefferson thought that the instruction of young children should consist, not in abstract moral precepts taught by a "platonized" Christianity, but in the aesthetically appealing moral examples of philosophical history.[67] As he regarded man in the Aristotelian tradition as an "imitative animal," and imitation as "the germ of all education,"[68] young children, before learning to employ their reason, could already learn to act morally simply by imitating the moral examples that their historical reading had impressed on their senses and memories.

Thus, Jefferson was able to think of even a relatively complex historical argument like that of the *Summary View* as a simple moral lesson that did not mainly involve the intellect. As he explained to the king, "The great principles of right and wrong are legible to every reader; to pursue them requires not the aid of many councellors."[69] (The fact that George III failed to learn what Jefferson regarded as an intellectually unchallenging insight accessible even to young children allowed Jefferson later to express contempt not so much for the king's exceptional wickedness as for his exceptional stupidity.)[70] The moral utility of philosophical history consisted, not in its appeal to the student's critical reason, but in the mechanistic imprint made by a historical incident or personality on the student's memory and his senses, including the moral sense. By a largely unreflecting reception of moral narratives about the past, the student was supposed to be able to train the physical strength of his moral sense for future use as he would that of "any particular limb of the body," without having to develop his rational faculties as well.

As Jefferson summed up his educational views in an oft-quoted letter to Peter Carr, a moral case could be decided "as well, & often better" by a ploughman than by a professor, since a professor, trained to rely on his reason, might be "led astray by artificial rules."[71] Since Jefferson regarded history as a collection of "moral cases" and historical studies as "moral exercise,"[72] he thought that it was intellectually accessible to "the great mass of the people," no matter what their level of education. The similar status that practical morality and philosophical history had in Jefferson's mind is illustrated by his statements about them as university subjects: in his educational letters to young men in the 1780s, he described attending university lectures in moral philosophy and history as, respectively, "lost time" and a "waste of time."[73]

This anti-intellectualist and anti-academic strain in Jefferson's early conception of history is most evident in a 1786 letter to his future son-in-law, Thomas Mann Randolph. After telling him not to waste his time with history professors but to acquire his historical knowledge solely "from books," he emphasized that this solitary historical reading should only "fill up those chasms of time not otherwise appropriated." Compared with the natural sciences, historical studies clearly should not be the priority of Randolph's Edinburgh education. Although Jefferson implicitly displayed another priority by dedicating the far greater part of the letter to history, he recommended this subject merely as a subordinate "after dinner" activity:

> There are portions of the day too when the mind should be eased, particularly after dinner it should be applied to lighter occupation: history is of this kind. It exercises principally the memory. Reflection also indeed is necessary but not generally in a laborious degree. To conduct yourself in this branch of science you have only to consider what æreas of it merit a grasp & what a particular attention, & in each ærea also to distinguish between the countries the knowledge of whose history will be useful & those where it suffices only to be not altogether ignorant. Having laid down your plan as to the branches of history you would pursue, the order of time will be your sufficient guide.[74]

In agreement with his understanding of history as accessible to any member of "the great mass," Jefferson considered it as a decidedly "lighter occupation" for a young man of the Virginia elite receiving a university education. Even though Randolph had the advantage of studying in a center of modern historiography, historical studies were merely pleasant–capable of "easing" the mind at night–as well as morally useful; they had little intellectual prestige for Jefferson, since they did not involve the faculty of reason "in a laborious degree."

A closer look at his references to time in this letter reveals that while

Jefferson regarded the linearity of rational time as capable of establishing "order" in Randolph's studies–"the order of time" was to structure both his daily schedule and the sequence of his historical reading–he stressed that the points on this time line were not all of equal importance in Randolph's personal development or in the content of his historical reading, the development of mankind. Some "portions of the day," for example, the twilight hours after dinner, had to be used less intensively than others. And some periods of history, for instance, the dark times of the Middle Ages, merited less attention than others. As was characteristic of philosophical history, Jefferson distinguished between historical periods according to their value to the present and future. This often meant that the study of periods of the past that were not immediately "useful" to eighteenth-century readers could be neglected. Thus, Jefferson's letter to Randolph implied a condescension toward "history" not only concerning the intellectual faculties required for its study but also concerning the value of its content.

In philosophical history, these two aspects often overlapped, and disparaging remarks directed at the intellectual requirements for *history* as a subject of study often merged with a derogatory attitude toward *history* referring to the past as such, or at least to some periods in the past. Expressions of enlightened contempt for the past, of course, were often shaped by the "modern" position in the seventeenth- and early eighteenth-century debates on the relationship between antiquity and the present age. Especially among the French *philosophes,* progressive thinkers in the later eighteenth century such as Jean le Rond d'Alembert or Jefferson's friend the Marquis de Chastellux radicalized the position of previous moderns and made aggressive claims even for the necessity of "forgetting" instead of remembering the past.[75] To be sure, such expressions of disdain for the past should be understood less as serious arguments for a universal amnesia than as rhetorical strategies employed to emphasize the writer's faith in progress. Jefferson certainly did not go as far as to advocate the burning of all books containing "useless" historical facts at the end of each century. Nevertheless, his conception of moral and political progress occasionally led him to polemical statements expressing a considerable condescension toward the past. Especially in his nineteenth-century writings, after such promising events as his own presidency had in Jefferson's opinion proved that progress was possible, he could confidently make his famous remarks deprecating both the study of history and history itself. Often quoted later as examples of a typically "Jeffersonian," forward-looking attitude, his presidential and retirement writings offer a collection of wonderfully history-phobic statements, including Jefferson's claims to regard history as capable of telling only "what bad government is," his pride of "that nation whose silent course of happiness furnishes nothing for history to say," and his profession to like "the dreams of the future better than the histories of the past."[76]

Since many of Jefferson's writings nevertheless deal extensively with historical problems, the future-oriented pathos of these statements did not mean that he condemned the study of history in its entirety. His emphasis on the future needs to be seen as a crucial part of, rather than a contrast to, his concept of the past. The didactic aim of historical studies as a popular "moral exercise" aiming at future progress, however, ultimately made his philosophical history an unstable approach, based as it was on the paradox of a priority of the future in a study of the past. Measured by the future-oriented standard of natural philosophy, history was frequently portrayed as an inferior field of study even by its proponents, who, like Bolingbroke, often argued more from defensive than from clearly affirmative positions. In an epistemology that understood "chiefly historical" learning not as a matter of reasoning but mainly as the result of a mechanistic imprint on the senses and the memory, a study of the past appeared as a mental operation that was all the more primitive for referring to sense impressions gained from a reality that was inherently stationary: unlike the future, the past by definition could not be changed and improved. For the more radical progressive thinkers in the later eighteenth century, knowledge of the past therefore often became a *quantité négligeable,* if not altogether a dangerous territory. Hence, what was often the content of philosophical history–a narrative of human progress–tended to compromise its didactic promise: if human knowledge was progressive, how could the past be expected to teach future generations anything at all?

An answer that allowed eighteenth-century thinkers like Jefferson to reconcile their faith in scientific, moral, and political progress with a simultaneous conviction of the instructive function of historical studies was to conceive of philosophical history as teaching negative lessons.[77] From his early writings on, Jefferson tended to imply that the lessons taught by history were mainly lessons from the past of what to avoid in the future. This tendency was already visible in the *Summary View,* although its long-term argument was framed by a traditional "restoration," and Jefferson still depicted European and American history as a more or less homogeneous whole. Yet in the short-term, Jefferson already admonished the king to break with the continuity of the recent past by refusing to imitate the actions of his immediate predecessors.[78] After the Declaration of Independence, his tendency to think of history as a collection of negative examples, different from what he hoped would happen in the present and future, increased (especially as this collection was from then on enlarged by the negative example of George III). With the success of the American Revolution, Jefferson sought a solution to the conundrum of philosophical history as a future-oriented study of the past. He began to differentiate more and more clearly between European and American history, directing his enlightened contempt for the past mainly at the former and trying to express some optimism, or at least remain undecided, in regard to the latter. He began to have hope–a hope that his nineteenth-century writings would often declare had been substantiated–that American history might become entirely different from the European past. In order to achieve this end, Virginia citizens had to learn not to emulate European history but to avoid its mistakes. In other words, the "moral exercise" Jefferson expected Virginians to gain from the study of history in the 1780s was an exercise, first of all, in the moral rejection of a past that he would identify, with increasing consistency, as un-American.

In Jefferson's middle and later writings, his tendency to associate different modes of time perception with different positions in space would enable him to differentiate between European and American history and find new reasons for studying the past of a future-oriented nation. Eventually, he developed his own mode of coming to terms with the paradox of philosophical history–of the didactic legitimation of retrospective studies in a progressive worldview. Yet the original negativity of a historical "moral exercise" continued to pose problems in his historical thought. For a progressive thinker like Jefferson, who saw the primary function of history in its ability to teach negative lessons of what to avoid, historical writing was likely to be considered inferior or at least incomplete if there was another form of writing that also managed to provide positive moral examples that could be emulated and transferred directly into active life. According to the Jefferson of the 1770s and 1780s, such a collection of positive moral examples existed outside of history. His opinion in this matter would change in his later years, but the younger Jefferson still thought that a positive "moral exercise" could be gained at least as well from the reading of fictional as from the reading of historical texts.

The Magic of Writing and the Mechanism of Reading

A characteristic of philosophical history that played a crucial role in shaping Jefferson's frame of mind was the Ciceronian subordination of historical narratives to rhetoric and their approximation to the pole of poetry.[79] The problem with history as "a moral exercise," Jefferson wrote to his future brother-in-law Robert Skipwith in 1771, was that "her lessons would be too infrequent if confined to real life. Of those recorded by historians few incidents have been attended with such circumstances as to excite in any high degree this sympathetic emotion of virtue. We are therefore wisely framed to be as warmly interested for a fictitious as for a real personage. The field of imagination is thus laid open to our use and lessons may be formed to illustrate and carry home to the heart every moral rule of life."[80] The young Jefferson considered the didactic function of historical writing to be precarious not because, as might be expected today (and as he would begin to argue later), he was afraid that it might deviate from past reality but precisely because he thought it was by definition too close to that reality. In order to be morally instructive, the above letter suggests, it would almost be preferable for historical writings *not* to be "confined to real life" but rather to partake, to some degree, in a "fictitious" world containing better illustrations of "moral rules." As they could more easily transcend the limits of a contingent factuality, "fictitious" events could be designed to impress on the moral sense "every moral rule of life." While historical and fictional writings differed in their means and the intensity of their effect, their larger aim was the same: their narratives should always allow the reader to "carry home to the heart" universal lessons in morality.

Due to this primacy of the moral function in both historiography and poetry, Jefferson did not see a fundamental difference in the ways historical and fictional writings represented the idealized nature of an ordered universe. Like many of his contemporaries, he did not tend to conceive of history and literature as different realms with separate epistemologies, requiring specific forms of historical-critical or fictional reading and writing. Instead, as the negative expression of confinement emphasizes, the young Jefferson even considered it rather impractical that historical lessons were "confined to real life." The standard by which to measure historiography and poetry was not faithfulness to a historical reality in one case or imaginative originality in the other. Because of their shared objective of teaching moral lessons, historiography and poetry appeared as two didactic literary genres that could be distinguished mainly by the degree of their salutary effect on the reader. In his letter to Skipwith, Jefferson argued for the relative advantages fictional texts had over historical ones, not so much in a qualitative as in a quantitative sense: he described in what number and the "degree" to which fictional more than historical texts were able to "excite . . . this sympathetic emotion of virtue."

Such a gradualist, rather than essentialist, understanding of the differences between history and poetry was in many ways characteristic of philosophical history. Historical writing appeared less to reflect a particular mental attitude toward the representation of past reality than to represent just one didactic literary genre among others. Thus, David Hume concluded that the composition of history and epic poetry differed "not in kind, but in degree." He could do so because his three "connecting principles" in fictional and historical productions, summarily referred to as "all compositions of genius" and "narrative compositions," were essentially the same as those in the association of ideas in the mind generally.[81] Historical as well as fictional writings reflected what he thought were essentially universal and unchanging principles of the mind.[82] As Hume saw the difference between history and poetry merely in different "degrees of connexion" within a universal concept of the mind, instead of a specific mode of perception unique to each form of writing at a given moment, he could make a statement that illustrates the differences between philosophical history and more modern approaches: "The PELOPONNESIAN war is a proper subject for history, the siege of ATHENS for an epic poem, and the death of ALCIBIADES for a tragedy." Deciding on the appropriate genre, according to Hume, was "a matter of taste more than of reasoning."[83]

This literary concept of history would be considered counterintuitive later on (until its renaissance in twentieth-century historical theories).[84] In a rationally grounded distinction between history and literature that would emerge fully with the transcendental turn in historiography,[85] each of the three events in Hume's example might become the subject matter for either historiography or poetry; it depended on one's particular mental attitude toward past reality. Contrary to this, the gradualist conception of philosophical history, as Hume theorized it, merely quantified different degrees of intensity within a universal reception process. In this vein, the philosophical historians distinguished between different qualities in the empirical subject matter instead, considering some topics in themselves better suited than others to achieve a certain (moral) effect. For them, the distinction between history and literature tended to depend on factors outside, not inside, the human mind. The young Jefferson of the letter to Skipwith could accordingly differentiate between outside events that merited closer study and those that did not, while caring little whether the narratives about them were real or imagined: "the fictitious murther of Duncan by Macbeth in Shakespeare" was able to have the same moral effect on the reader's senses "as the real one of Henry IV. by Ravaillac as related by Davila."[86]

This literary conception of history was negatively defined by an Aristotelian argument that it was seeking to transcend. In the *Poetics* Aristotle had made a crucial distinction between historiography and poetry. He had characterized poetry as the more serious and more philosophical of the two, since the poet need not confine himself to particular incidents that had accidentally happened, but was supposed to describe incidents that could happen "probably or necessarily."[87] Jefferson's early approach to history was transitional. Following Aristotle's interpreters since the Renaissance, he acknowledged a conception of mimesis as the imitation of the probable, attributing relatively more value to texts about events that were likely or supposed to happen than to historical accounts of events that had merely chanced to happen. In the tradition of Aristotle's estimate of poetry as more philosophical than historiography, literary narratives about "fictitious" events were, in a sense, the better philosophical history for Jefferson at this point, as they were better equipped to convey the necessary "moral rules." However, Jefferson also partook in the tendency popularized by Bolingbroke and theorized by Hume to undermine the Aristotelian classification of historiography and poetry. By emphasizing the shared function of both forms of writing, he tended to level the conventional hierarchy between "noble" genres like epic poetry or tragedy and "plain" historical records. Identified with the poetic function of dealing with the probable, philosophical history thus gained in status. By claiming the traditionally poetic territory of telling philosophical rather than accidental truths, it became one of the literary parvenus of the eighteenth century.

Another hopeful parvenu in the literary system of the Enlightenment was, of course, the relatively new epic genre of the novel. In order to understand the literary classification of philosophical history that shaped the young Jefferson's thinking, it is useful to briefly consider that there was a mutual approximation between both literary newcomers in the eighteenth century, before they began to drift apart and writers began to construct scientifically historical or self-confidently fictional (rather than merely "fictitious") narratives. In the strategies of legitimization they employed before this professionalized distinction, novels and historiographical writings tended to overlap and sometimes, rather paradoxically, even surpass each other in opposite directions. Eighteenth-century novels, on the one hand, tried to partake in the rise of philosophical history, not only by stressing their shared didactic aim in describing what was likely to happen but even by the fiction of describing what had actually happened. By presenting their novels as collections of historical documents (letters, diaries, etc.) and using the term *history* in their titles,[88] many novelists aspired to a more genuinely "historical" subject matter than many of the contemporary historians themselves, on the other hand, would have liked to be identified with. As echoed in Jefferson's Skipwith letter, it was a characteristic tendency in philosophical history, on the contrary, *not* to be proud of a particular closeness to "real life," but instead to criticize the "confinement" of historical writings to the contingency of the past.

Enlightened theorists of history thus accepted, to some extent, that their priority of partaking in the literary domain of conveying universal moral lessons could be at the expense of factual details. Directing their enlightened contempt for the past at the antiquarian historians of previous generations,[89] they even prided themselves on the fact that their works consisted in useful generalizations rather than accurate historical records. As Bolingbroke arrogantly said: "I had rather take the Darius whom Alexander conquered, for the son of Hystaspes, and make as many anachronisms as a Jewish chronologer, than sacrifice half my life to collect all the learned lumber that fills the head of an antiquary." In his letter to Skipwith the young Jefferson quoted Bolingbroke's comment approvingly and ridiculed pedants who filled their heads with "the learned lumber of Greek and Roman reading."[90] Bolingbroke's criticism of older, less literary forms of professional historiography–annals and chronology–may have appealed to Jefferson especially because these approaches tended to be based on views of the past that were structured by dynastic and institutionalized religious events.

Jefferson also followed Bolingbroke in the area in which the latter may appear most modern today, namely, in his "historical" reading of the Bible, which neglected its traditionally privileged status as religious revelation. As the early entries from Jefferson's *Literary Commonplace Book* suggest, he probably agreed with Bolingbroke that it was necessary to give the same credit to Moses "as we give to every other historian."[91] Bolingbroke's new historical standards were often literary standards as well, as when he advised to examine historical periods "as you would examine a tragedy or a comedy."[92] Jefferson affirmed Bolingbroke's historical-literary standards when he advised his nephew Peter Carr to read the Bible, including apocryphal texts, as a stylistically refined history: "Read the bible then, as you would read Livy or Tacitus." Livy especially represented a form of historical writing that gave priority to literary presentation rather than to factual precision–in this sense an ancient precursor to the poetic aspirations of philosophical history. Jefferson commonplaced Bolingbroke for a comparison between the "incredible anecdotes" in Livy and the Old Testament, measuring both by a universal standard of probability. The "historical" treatment of scriptural texts that Jefferson learned from Bolingbroke's comparison of the Bible to Livy is therefore difficult to distinguish from a procedure of neoclassical literary criticism: oriented toward an Aristotelian idea of poetry as the imitation of the probable, Bolingbroke and the young Jefferson criticized Livy's historical work and the Bible for their "improbable" historical anecdotes. Instead of trying to explain these anecdotes as part of a particular religious or intellectual context, they tended to reject them altogether for their mimetic deficiencies. In the final analysis, then, they criticized these works not so much for their historical inaccuracies as for their literary inferiority.[93]

The boundaries between historical, poetic, and sacred writing thus being blurred, Jefferson would variously recommend religious, fictional, and historical works for moral instruction in the 1770s and 1780s. While the sections dedicated to purely literary works in his later reading lists did not attain the dimensions of the list attached to the Skipwith letter, Jefferson would still praise Sterne's sermons and novels as "the best course of morality that ever was written" in 1787.[94] He also liked to recommend works such as Goldsmith's histories or Fénélon's *Télémaque* (published in 1699), which merged fictional and historical elements.[95] Throughout his life, moreover, Jefferson remained fond of reading ancient authors in the original, recommending the preference of "original authors" to "compilers" to Peter Carr, Thomas Mann Randolph, and his daughter Martha.[96] Jefferson's preference for original authors may sometimes have reflected a sense of historical authenticity,[97] while at other times it appeared primarily grounded in his admiration for their literary qualities. From his perspective, a decision whether to privilege their historical, their political, or their literary qualities would have appeared out of place: he could read ancient writings as part of a particular historical reality, as examples of universal political problems, and as timeless stylistic models.[98]

This attitude was reflected in Jefferson's lack of systematic distinction between primary and secondary literature.[99] He could be rather indiscriminate in listing ancient and modern authors under the same heading, writing that Peter Carr's reading in "Roman history" should consist in "Livy, Sallust, Caesar, Cicero's epistles, Suetonius, Tacitus, Gibbon."[100] In this context, the primary meaning of "Roman history" was not the Roman past as such; in this case Jefferson might be expected to have listed Gibbon separately, who wrote about, but obviously did not belong to, the past of Rome. What Jefferson meant by "Roman history" here was the writing about the Roman past, something like the genre of *all* writing relating to ancient Rome, modern and ancient histories as well as Cicero's letters. Jefferson's use of the term *history* thus referred to a transhistorical literary genre defined less by the individual consciousness of the writer than by the writer's subject matter.[101]

This sense is often present in Jefferson's use of the term *history* when it referred exclusively to the genre of historiography, the writing about past events, rather than summarily to the past events themselves. The "chapters" and "pages" of "history" mentioned by Jefferson were still closer, in this sense, to the actual ink and paper of historical writing than the same expressions would be today. Referring to a literal "page" of history, Jefferson could even use *history* and *truth* as contrary terms.[102] *History* understood as historical writing approached nearer to *art* in two meanings of the term. Its negative connotations would become prominent in Jefferson's later years, when he began to fear that historical works might become the medium by which dangerous political ideas were "artfully insinuated."[103] A different sense of *art* shaped Jefferson's understanding of history from early on. When he described historical studies as "a moral exercise," he subordinated historical works to the classical art of rhetoric. In this context, *history* was less a philosophical term referring to the past as such than mostly a technical term referring to a collection of examples supposed to influence an audience.[104]

Jefferson illustrated this rhetorical function of history by references to the art of painting. Like many of his contemporaries, he typically imagined historians or biographers to be in the process of "painting," giving "lustre," by more or less "delicate touches," to "portraits" on the historical "canvas."[105] Such a metaphorical identification of historiography with painting relied on the assumption of the common mimetic function of poetry and the visual arts condensed in the formula *Ut pictura poesis* from Horace's *Ars poetica.* Although eighteenth-century thinkers (including, eventually, Jefferson himself) began to stress qualitative differences between the two forms of art,[106] his early conception of history shared the static character of the visual arts. Metaphorically speaking, Jefferson's philosophical history was not the forceful "stream" of history described by later writers,[107] but came closer to the "confined" space of an old archive that contained a mass of useless materials and a limited collection of satisfactory historical paintings.

Jefferson's identification of history with art implied a premodern concept of art, one based not on uniqueness and originality but on craftsmanship and the mastery of a canon of rules required for the imitation of an ordered universe. According to this universalist emphasis, he tended to measure historiography by criteria that had to do more with the composition of the narrative than with the examination of the narrated past. These criteria, concerning mainly the selection of material and its artistic presentation, aimed at maximizing the moral effect on the reader. This effect depended on the "design" of the narrative in a double sense: the intention of the author and the mastery of his art.[108] Jefferson imagined that a good historian "designed" his narrative by actively selecting an enlightening subject matter and passing the correct moral judgments. Like an accomplished orator or painter, the historian had to be in control of his medium, because the moral effect of his work, as Jefferson frequently emphasized, depended heavily on the refinement of his literary "style."

A sensitive stylist in his own writings, Jefferson was scrupulous in distinguishing between the styles of the authors he read. He cherished, for instance, the fictional speeches in Livy's histories for their style, while he was uncommonly critical of those of Cicero, whose style in general he nevertheless thought to reveal "the first master in the world." Among modern writers, Jefferson appreciated, for example, Bolingbroke's neoclassicism as "a style of the highest order: the lofty, rhythmical, full-flowing eloquence of Cicero."[109] In his summary, in *Notes on Virginia,* of the historical works written on his native state, Jefferson seems almost to have been more interested in their stylistic qualities than in their factual precision. For instance, he criticized William Stith's *History of the First Discovery and Settlement of Virginia* (1747) along the lines of Bolingbroke's "learned lumber" criticism: Stith, according to Jefferson, was "very exact, but of no taste in style. He is inelegant, therefore, and his details often too minute to be tolerable, even to a native of the country, whose history he writes."[110]

A brilliant style, however, could also become a problem. According to the neoclassical maxim that art consisted in concealing art–*Ars est celare artem*–there was a sinister side to artists who had mastered their medium sufficiently to be able to disguise their artistic mastery. Especially in an epistemology with a nominalist attitude toward language, reading well-crafted literary works might become a dangerous activity, as the reader could never be sure of the writer's true "design." Reflecting on historiography and fiction, the young Jefferson described this problem optimistically: "If the painting be lively, and a tolerable picture of nature, we are thrown into a reverie, from which if we awaken it is the fault of the writer."[111] Only bad writers painted nature so inadequately as to destroy the illusion of their readers, whereas masterly writers managed to blur the boundaries between art and life, fiction and fact, influencing their readers by the trompe l'oeil naturalness of their style. Thus, it was essentially a self-controlled duplicity, a paradoxical ability to conceal an artful lack of naturalness through art, that had the hypnotic effect of actively "throwing" passive readers into "reveries."

For a generation who considered reading mainly as "a moral exercise," this supposed lack of sincerity underlying the success of influential writers also had a political dimension. As John Adams wrote to Jefferson, "Style has governed the Empire. Swift, Pope and Hume have disgraced all the honest Historians. Rapin and Burnet Oldmixen and Coke, contain more honest Truth than Hume and Clarendon and all their disciples and Imitators. But Who reads any of them at this day?"[112] A self-controlled literary style could appear an important source of political power: not only for Adams, a masterful style seemed able to "govern" the empire more effectively than plain (and boring) historical honesty. According to the aged Jefferson's parallel judgment of Hume, style could even be a more influential political factor than "all the million of men in arms of Bonaparte."[113] Perhaps due to the indirectness of its influence, Enlightenment thinkers sometimes went as far as to identify the effects of a brilliant style with "magic." Like Gibbon, who criticized Voltaire for using "the magic of his style" to conceal the factual imprecisions of his historical *Essai sur les mœurs* (1756),[114] Jefferson would criticize Hume for abusing "the magic of his art" to convert innocent young readers into Tories.[115] Hume's *History of England* would cause the nineteenth-century Jefferson his greatest worries about the dangerous effects of a "bewitching" style in historiography. Yet his fear that "master magicians" might abuse their stylistic powers to recruit political converts on the wrong side had its roots in a basic asymmetry characterizing his understanding of literature from early on.[116]

In a transitional phase in the eighteenth century when an aesthetics of effect was beginning to replace a traditional canon of rules, Jefferson's emphasis on stylistic mastery made him exaggerate the powers of the writer at the expense of the reader. In order to disguise his design and paint "lively" pictures of nature, the master writer himself had to know the difference between nature and art, fact and fiction, very well. The reader, on the other hand, had no means to acquire this information, especially if the work of fiction or history had been done by a "master magician." As the young Jefferson (apparently still unaware of the problematic implications of this asymmetry) cheerfully explained to Skipwith, readers were passive, undiscriminating recipients of fictional and historical writings, who generally "neither know nor care" or "reflect whether the story . . . be truth or fiction."[117] Understanding historical and fictional reading as "a moral exercise," the young Jefferson did not think that it required a great deal of rational reflection. The "reader of feeling and sentiment," as he significantly put it, had to rely on his sense impressions and "emotions."[118]

In keeping with the use of painting metaphors for historical works, eighteenth-century theorists tended to privilege the sense of sight in the "moral exercise" of history. For example, in *Elements of Criticism* (1762), which was on Jefferson's reading list for Skipwith, Lord Kames had explained the close connection between sight and moral action, claiming that "even real events, entitled to our belief" had to be presented "as passing before our sight" in order to "move" the reader and achieve a moral effect. Thus, readers were turned into "spectators."[119] Jefferson followed Kames in his assumption of an especially close connection between moral instruction and the sense of sight. Like Adams, whose image of a mirrored boudoir emphasized the visual character of philosophical history, Jefferson imagined, for instance, a historical argument like that of the *Summary View of the Rights of British America* to offer its reader a "view" of a political lesson.[120] As he saw it, fictional as well as historical writings were able, as paintings were, spontaneously to "illustrate and carry home to the heart every moral rule of life." Whereas Jefferson conceived of the writing process as a rationally mediated act by master writers, he understood the reading process mainly as an involuntary process of sensory perception. Overstating the passive enjoyment of reading, Jefferson's metaphors could even go further than emphasizing the sense of sight, identifying the practice of reading with purely physical processes involving food, drink, or sexuality: after the writer had actively "instilled," "inculcated," or "infiltrated" the reader with a certain moral content, the reader would automatically be "nourished," "poisoned," or "seduced."[121]

The conception of the reader as a passive consumer of art had far-reaching implications in a universe where similar causes were expected to produce similar effects and in a literary system based on mimesis as the imitation of the probable. Master writers and spectator-readers in this frame of reference were linked by a "chain of causes and effects" that tended to be loosened in the nineteenth and twentieth centuries, when the expectation of a mimetic relationship between intention, writing, reading, and moral action would eventually disappear. In philosophical history, by contrast, writers were ideally expected to imitate the probable for readers, who imitated the content of their reading by their actions. Jefferson's concept of moral and political education presupposed a passive reader–the Aristotelian "imitative animal" he mentioned in *Notes on Virginia*[122]–who automatically formed habits copying the vices or virtues of his (fictional or historical) reading. Since reason was not assumed to play a crucial role in this mechanism, the writer's intentions would eventually be reflected in the actions of the reader, who was not expected to rationally decide whether he wanted to "copy the fair example" or not.[123] While the writer was an independent agent, the reader tended to be an unreflecting link in a "chain of causes and effects" designed by the writer. Ultimately, in this train of thought, the writer was the master, and the reader merely his tool or–an interchangeable identity, according to Aristotle–his slave.[124]

In Jefferson's early years, as documented by the Skipwith letter, this problematic implication of a literary historiography rarely came to the surface. Before "making" history himself, the young Jefferson could apparently retain his peace of mind when confronted with stylistic masters who willfully distorted the facts–or even prefer them altogether to the "learned lumber" of antiquarian historians. The superior style of "master magicians" promised to be both more entertaining and morally more effective than the dry accumulation of historical details. Bolingbroke's "magic of example" and the literary procedure of selecting and skillfully presenting timeless moral examples appeared to be the most legitimate use of history. Beginning in the revolutionary years, however, Jefferson experienced new situations in which the literary presentation of historical examples turned out to be a risky business. As a consequence, his concept of history as "philosophy teaching by examples" became more complex. In his praise of the great rhetorical use of examples in historiography, Bolingbroke had anticipated a problem that would seriously alarm Jefferson in his middle and later years: "The harshest orders are softened by example, and tyranny itself becomes persuasive."[125]

Chapter 5

Seduction by Example

The final two decades of the eighteenth century were a period of great personal and political turmoil for Jefferson, challenging his enlightened conception of a teleological history. The 1780s began with his troubled wartime governorship of Virginia and Tarleton's raid on Monticello. They were darkened especially by his wife's early death in 1782–a catastrophic "single event" that "wiped away" all Jefferson's plans for "future happiness."[1] This decade also produced the greatest spatial dislocation in his life: his years as American commissioner and minister to France, from 1784 to 1789, not only brought him into direct contact with the historical particularities of different European societies but also made him an eyewitness to the cultural, social, and political "fermentation" of the beginning French Revolution.[2]

Beginning with his return to America, the 1790s were hardly a decade conducive to stabilizing Jefferson's conception of a providential historical order. As secretary of state, vice president, and increasingly a party leader, he became personally involved in what he and many contemporaries experienced as a great epistemological crisis: the confusion of "suspicions & certainties, rumors & realities, facts & falsehoods, . . . principles & pretensions"[3] in the battles of an emerging party politics. In this context, an important assumption underlying much eighteenth-century history writing–the expectation of a moral homogeneity of cause and effect, of a largely harmonious relation between motivation, intention, and action, and of a consequent integrity of "character"[4]–often appeared incompatible with political experience. This trend was aggravated by the radical turn of the French Revolution, which proved to many at the time that the most evil effects could follow from seemingly benevolent causes. To an initial supporter such as Jefferson, who had agreed, in 1789, "to be stoned as a false prophet if all does not end well in this country,"[5] and who had still defended the French Revolution at the beginning of 1793,[6] this insight must have been rather unsettling (even without the personal accusations of Jacobinism in the Federalist press). The Terror of the French Revolution and the racial wars in the former French colony of Saint Domingue further complicated the didactic function of history, and the "magic of example" that the young Jefferson had optimistically admired in Bolingbroke increasingly displayed its dark sides. In both domestic and foreign policy, universalist interpretations not only of ancient but also of modern history could have threatening implications. If slaves and free mulattoes in Saint Domingue, for instance, were successfully (and, as Jefferson realized, legitimately) shaking off the chains of their white oppressors, why should American slaves and free blacks not try to follow their historical example? And if the French had partly modeled their revolution on the example of the American Revolution,[7] what could it mean for the American original that its French copy had failed? Ultimately, such questions contributed to undermining the universalism of philosophical history. Progressives like Jefferson were particularly under pressure to question homogenizing conceptions of historical time: they had to try hard to envision a "course of human events" in the young United States that was decidedly different from events in France and its colonies.

Jefferson's personal experience of the revolutionary period thus functioned as a catalyst in the transformation of his historical thought. However, he had to find means to loosen the historical "chain of causes and effects" also for reasons that went beyond the problems of contemporary politics. As discussed in the previous chapter, Jefferson's philosophical history was an unstable concept that was prone to being undermined by its own conflicting solutions to two basic problems: the moral status of historical facts and the nature of historical time. The first problem concerned the relationship between writers, readers, and agents of history. Although conceived as "a moral exercise," philosophical history was understood to achieve the most powerful effect if written by "insincere" stylistic magicians with a questionable morality. Throughout the eighteenth century, it had suffered from this moral paradox and the potentially troubled relation between the priority of its didactic function and the accuracy of the historical facts and details it professed to describe. Understood as a literary genre and a mimetic "art," it was an easy prey to the pessimistic implications of eighteenth-century empiricism. Meant to teach moral lessons by the artistic presentation of moral examples appealing, in particular, to the sense of sight, philosophical history presupposed passive, unreflecting readers who were easily deceived by the trompe l'oeil effect of a well-crafted neoclassical style. Like the mirror images of the young lady in Adams's historical "boudoir," as the contemporaries began to realize, the series of quasi-visual impressions to be gained from the reading of philosophical history was ultimately tautological: it always threatened to consist in nothing but slightly distorted reflections of the eighteenth-century spectators themselves.

And like the mirror images in Adams's "boudoir," the examples of philosophical history had a very limited capacity to account for temporal change. As they were essentially self-reflective, they were capable of depicting at best some quantitative, but not much qualitative, change in time. Apart from its problematic character as a literary "art," philosophical history also suffered from its temporal paradox of giving a priority to the future in a study of the past. Founded on the assumption of transhistorical laws of (human) nature, it was meant to achieve an aim that was modeled by a "false analogy" on the natural sciences: the aim of predicting the future based on past experience. Illustrated by the static spatial metaphors of a rigid chain, a mimetic painting, or a mirrored "boudoir," philosophical history had elementary difficulties with integrating the dramatic "acceleration" of change that Jefferson, like many of his contemporaries, observed in the late eighteenth century. With the intrusion of a qualitative concept of change that depended on contingent human actions and perceptions, the homogeneous metaphorical space of philosophical history threatened to collapse.

Jefferson occasionally professed to be overwhelmed by the paradoxes of philosophical history and the new historical reality that brought them to the surface. For instance, he justified his second retirement, from 1794 to 1796, by the claim of having become physically unable to cope with the rapid changes and disorder in contemporary historical events: "The motion of my blood no longer keeps time with the tumult of the world."[8] This sentence established a contrast crucial to a philosophical trend initiated in the earlier eighteenth century (although without much contemporary influence) by Giambattista Vico, the contrast between natural and civil history that would become decisive for what Isaiah Berlin would call the "Counter-Enlightenment" and the emergence of a modern historical thinking.[9] Jefferson located this contrast between an individual movement subject to the steady rhythm of a cyclical physical nature (the regular motion of his blood) and a collective movement subjected to the chaos of human history (the "tumult of the world"). He implied that the contrast was new to him: apparently, there had been a period in the past when the two movements had still "kept time," when the rational time of natural laws and the sentimental time of human history had formed a harmonious whole.

Jefferson's impression of being unable to "keep time" with a tumultuous revolutionary world suggests a heightened awareness that this new reality began to require new modes of historical explanation. In his early years, physical nature and civil history had still appeared to converge in their equal subjection to rational time. Yet in the crises of his later life, they began metaphorically to drift apart, as if to corroborate Vico's assumption of two distinct realms for nature and history. In the process, Jefferson's youthful conviction of a homogeneous "chain of causes and effects" was stretched almost to its breaking point. His writings from the crisis-prone years in the late eighteenth century thus contain his most historist or protohistoricist statements.[10] As in other domains, however, Jefferson's eclectic thought did not tend toward a systematic analysis of this development. Stopping short of a categorical break with philosophical history, the transformation of his historical consciousness was mainly reflected in negative statements expressing disillusionment with or skepticism about the historical practices he encountered.

Jefferson had a favorite metaphor for the destabilizing expectation of radical change in his *fin-de-siècle* crisis. Like many of his contemporaries, he condensed the stressful experience of this period in the image of a "storm" or a "hurricane." In an enlightened understanding of nature, storms were devastating, but they also had their limits and their necessity. They represented an extreme of disorder, arbitrary change, and acceleration of movement, but their transitory nature could also promise to restore a previous equilibrium.[11] Thus, the natural metaphor of the storm indicated how modern Jefferson's historical thought could become, how far the framework of his early conception of natural laws in history allowed him to go in his middle and later years. This chapter explores the transformations of Jefferson's historical thought when the global revolutionary "hurricane" of the late eighteenth century threatened to make landfall in America.

A Fundamentally Different Past

In Jefferson's early understanding of philosophical history, the factual status of a moral example had been relatively inconsequential, as it had almost seemed more important for a historical narrative to be instructive for the future than to describe what had actually happened in the past. In the last quarter of the eighteenth century, however, Jefferson lost his nonchalance about the interchangeability of fact and fiction in historical works. He even became rather obsessed with the distinction of "fact and fable" or "truth and falsehood" in historiography and remained so until his retirement. While he was in the process of becoming aware of "making" history himself, the question of historical accuracy–ultimately, the Romantic question of how it actually was, of what had "really" happened–was becoming a more personal problem. Jefferson's explicit concern with historical factuality may have commenced with his draft of the Declaration of Independence. In a phrase that probably seemed exaggerated to his colleagues in Congress, he claimed absolute "truth" for his interpretation of Anglo-American history. Introducing the long list of violations of American rights by the king, he tried to emphasize the famous phrase "To prove this let facts be submitted to a candid world" by a relative clause that was eventually deleted in the congressional version: "for the truth of which we pledge a faith yet unsullied by falsehood."[12] By implying a contrast between British filth and American purity, Jefferson's draft identified not only a stereotypical innocence but also a specific truthfulness in historical facts as an exclusively American attribute. His formulation was somewhat unfortunate, as the temporal precision of "yet" ("yet unsullied by falsehood") left readers to wonder *when* he expected the purity of the American faith to be sullied by falsehood. Nevertheless, Jefferson's emphasis on the faithful presentation of the "facts submitted to a candid world" may have been the beginning of a new understanding of historical accuracy: after making a "truthful" adherence to historical factuality a national American characteristic, he had set the stage for a polemical struggle with historical facts in his middle and later years.

In Jefferson's writings from the 1780s, the exemplary function of history increasingly collided with this emphasis on the accuracy of historical facts. *Notes on the State of Virginia* presents an instance when Jefferson's initial optimism about the didactic value of historical examples was gradually diminished by an awareness of the complexity of past realities. In the first half of query 13, as has been described, Jefferson still stressed the universality of human nature, referred to a timeless "chain of causes and effects," and presented a collection of didactic examples from antiquity to the Renaissance. By the end of the query, however, he had discovered that the application of ancient examples to modern politics could be a highly problematic procedure. In his complex explanation of the "capital defects" of the Virginia Constitution of 1776, he passionately criticized two recent proposals of a dictatorship in the Virginia house of delegates during the Revolutionary War.[13] The political and personal intricacies of his argument aside,[14] the dictatorship issue also turned out to present a serious conceptual problem for Jefferson's historical thought. After he quoted a theoretical reflection on the problem of "translating" past examples into the present,[15] his misgivings grew more intense in the final pages of the query. Perhaps having in mind Bolingbroke's ambiguous praise of the "magic of example"–"The harshest orders are softened by example, and tyranny itself becomes persuasive"–Jefferson worried that even his well-meaning Virginia compatriots might be corrupted if they modeled their behavior on the ancient example of a wartime dictatorship:

> Those who meant well, of the advocates of this measure, . . . had been seduced in their judgment by the example of an ancient republic, whose constitution and circumstances were fundamentally different. They had sought this precedent in the history of Rome, where alone it was to be found, and where at length too it had proved fatal. They had taken it from a republic, rent by the most bitter factions and tumults, where the government was of a heavy-handed unfeeling aristocracy, over a people ferocious, and rendered desperate by poverty and wretchedness. . . . They misapplied this precedent to a people, mild in their dispositions, patient under their trial, united for the public liberty, and affectionate to their leaders.[16]

Jefferson followed three lines of argument against the "seduction" by an exemplary reading of ancient Roman history. He began by stressing the lack of legal continuity between the Roman precedents and the Virginian republic. Moreover, he emphasized that the content of the historical example was by itself dangerous, since the possibility of temporary dictatorships had also "proved fatal" in ancient Rome. The Romans had essentially been seduced by their own examples–in Jefferson's words, because their "temporary tyrant, after a few examples, became perpetual," and an initially well-meant example soon deteriorated into a "bad example."[17] This attitude reflected a trend in eighteenth-century historical thinking. To many progressive minds, ancient history was flawed as a model for modern politics, since its outcome was always negative: whether of the Greek polis, the Roman republic, or the Roman Empire, enlightened views on ancient history tended to be dominated by the theme of an eventual "decline and fall."

Jefferson's claim that temporary dictatorships had also led to political corruption in the past thus might have sufficed to counter arguments based on the Roman precedent. Yet this point was only directed against the emulation of historical examples; it still allowed for an enlightened reading that understood Roman history as a collection of negative lessons, teaching at least what to avoid in the future. Jefferson's skepticism about the didactic use of historical examples, however, went further than that. He radically argued that historical examples connected to the present were altogether "misapplied." Not only would their content be destructive if emulated but the method of transferring ancient examples to modern situations was itself erroneous, because, as Jefferson put it, historical conditions in ancient Rome had been "fundamentally different" from those of the present. The difference consisted in their respective stages in the moral development of civilization. The ancient Romans, as Jefferson characterized them, were "bitter," "unfeeling," "ferocious," and divided by "factions and tumults" into two opposed groups: a "heavy-handed" aristocracy and a poor, "desperate" people. Modern Virginians, by contrast, were morally and politically refined. "Mild," "patient," and "affectionate to their leaders," they constituted a homogeneous citizenry that was, as he liked to see it, "united for the public liberty."

Thus, the universality of human nature that Jefferson emphasized in the beginning of query 13 was significantly undermined in its final pages. He even claimed radical historical uniqueness for the past with regard to the problem of temporary dictatorships, contending that Roman history was "where *alone* it was to be found." Parallel to this stress on human particularities in time, the Jefferson of *Notes on Virginia* thus also began to emphasize particularities in space. A similar proto-Romantic (and protoracist) differentiation within the universality of human nature has made the following query, 14, notorious. There Jefferson claimed that the "difference" between people of European and African origin was "real" and "fixed in nature."[18] It was at this moment when he became occupied with seemingly unbridgeable differences between human beings in time and space that philosophical history became problematic as well: as he qualified the universalism of human nature, he also began to question, explicitly, the assumption of an exemplary use of history. And when the didactic function of history was in doubt, the particularities of past and present began to assume a value of their own.

To some extent, Jefferson's growing sensitivity to historical difference was typical of enlightened thinkers, who, more assertively than previous generations, began to identify with their particular century. This self-conception as contemporaries of "this eighteenth century"[19] was linked to a heightened awareness of anachronisms, especially in visual representations of history. If the past was "fundamentally different" from the present, this difference also had to be reflected in the visual attributes of historical characters. The last decades of the eighteenth century witnessed an especially strong emphasis on period costume, whether in historical dramas, paintings, or monuments. This emphasis was so prominent in the final years of the *ancien régime* that it found an ironic expression, for instance, in a historical fashion that mimicked the costumes and painting style of seventeenth-century Dutch masters. Jefferson appears to have found the historical playfulness of this early "retrofashion" amusing in its European context,[20] but in regard to the visual representation of American history his sense of irony was rather limited. In the 1780s he insisted on "the size of the life" and the "real costume, to wi(t), the military uniform" for statues of George Washington, professing to detest anachronistic details. As Jefferson told Washington, this dislike reflected "strongly the sentiment of West, Copley, Trumbull, and Brown, in London; after which, it would be ridiculous to add, that it was my own. I think a modern in an antique dress as just an object of ridicule as a Hercules or Marius with a periwig and a chapeau bras."[21]

This statement is especially interesting as Jefferson's taste in artistic representations of history would change significantly later in life: by 1816 he followed classical conventions in recommending colossal size and ancient costume for a statue of Washington.[22] In the late eighteenth century, by contrast, he was eager to give priority to the limits of the contemporary and historical, emphasizing the necessity for period costumes in the representation of both modern and ancient times. The reference to Marius in the letter to Washington quoted above may have been inspired by the celebrated painting *Marius prisonnier à Minturnes,* by Jean Germain Drouais, which he had seen in the Salon of 1787 earlier in the month.[23] In this work, as in Jacques-Louis David's *Le mort de Socrate,* which Jefferson also admired in this salon, ancient characters were depicted in attire inspired by antiquity.[24] Jefferson's emphasis on period costume was rather obsessive lest he or Washington should appear to a European public as "ridiculous" or "an object of ridicule" in the important domain of artistic taste. He seemed to fear that American characters depicted in classical costumes would suggest a lower stage in the progress of civilization–as if American taste had been arrested at a previous stage when historical characters were depicted in the dress of the artist's period–or an embarrassing ignorance of a past that Jefferson now regarded as "fundamentally different" from the present.

This respect for historical difference also shaped Jefferson's attitude toward narrative representations of history in the 1780s and 1790s. Explaining to his future son-in-law in 1786 why to read "original authors & not compilers," he argued that an "author who writes of his own times or of times near his own presents in his own ideas & manner the best picture of the moment of which he writes."[25] This statement suggests two things. On the one hand, it reveals that Jefferson still failed to draw a clear line between historians and their subject matter, merely distinguishing between modern "compilers" and ancient "original authors," whose writings were contemporaneous with the events they described.[26] Late in life he still evoked the classical ideal of a historical narrator with personal experience of his subject matter, for example, when he narrowly referred to Edward Gibbon, not as a brilliant historian of the Roman Empire, but merely as a modern "compiler."[27] In the distinction between "compiler" and "original author" the figure of the historian remained relatively pale. The historian capable of describing a unique reality by a particular historical method had not gained a sharp profile, requiring the supplementary attribute of contemporaneousness to be expected aptly to deal with his historical subject matter.

On the other hand, however, Jefferson began to realize how much depended on the historian's individual intellectual activity, on his "*own* ideas and manner." At least, it began to dawn on him that even writers contemporaneous to their subject matter were often unable to present anything but distorted "pictures of the moment" they claimed to describe. During his stay in Europe from 1784 to 1789, Jefferson began to be more immediately troubled by the question how recent American history would be represented by contemporary writers. In his concern with the historical representation of the American Revolution, he took active steps to influence what he thought would be the future shape of his period in historical memory. Whether in the case of David Ramsay's *History of the Revolution of South-Carolina* (1785), François Soulés's *Histoire des troubles de l'Amérique anglais* (1785), William Gordon's *History of the Rise, Progress, and Establishment of the Independence of the United States of America* (1788), or Jean-Nicolas Démeunier's articles in the *Encyclopédie méthodique* (1782–1832),

Jefferson made an effort not only to read the work closely but also to correspond with the author, whom he generously supplied with additional source material as well as comments on what he thought were factual errors in the texts.[28] Critically reading these works of history, he referred to them as significant contributions to "historical science."[29]

This science, however, often turned out to yield problematic results, especially, as Jefferson began to fear, when the research had been done by foreigners with limited knowledge of and, worse still, limited sympathies for the young United States, such as François Soulés. A few years previously, Jefferson had complained in *Notes on Virginia* about misrepresentations of America in the *Histoire des deux Indes* (first published in 1770) by the abbé Raynal and his collaborators. While these views had mainly concerned natural history (including Raynal's extended reference to the "natural" degeneration of white Americans), the errors that worried Jefferson in the later 1780s, in Démeunier's *Encyclopédie* articles and Soulés's *Histoire,* for the most part pertained to civil history alone, to the military and political history of the Revolutionary War and the discussions on slave emancipation.[30] By communicating his firsthand knowledge of the American Revolution, Jefferson hoped to present what he regarded as a more truthful picture to a European audience. In the case of Démeunier's *Encyclopédie,* Jefferson was particularly concerned lest the Virginia revolutionaries appear to have lacked "virtue & firmness" in their failure to emancipate their slaves.[31] However, he eventually had to realize that his efforts were rarely met with the success he had expected. Even his vast supply of trustworthy information did not automatically lead to a correct historiographical presentation of the historical "moment." Nor did it induce his contemporaries to relate the events in the appropriate "ideas and manner."[32] Jefferson made the depressing discovery that history could be rather a complicated business—even if historians had the advantage of direct access to as reliable a historical witness as Jefferson himself.

During his unsatisfactory exchanges with contemporary writers, Jefferson became acutely aware that the representation of historical facts in writing could be a serious problem in itself. As a result, he became skeptical toward historiography as a genre. This skepticism was considerably different from his youthful cheerfulness about the interchangeability of fact and fiction. In his middle years, Jefferson tended to emphasize the importance of historical details in historiography, while expressing serious doubts that historical reality could ever be successfully represented in writing. Even, for instance, in a letter to an American compatriot and friend who had had many of the same historical experiences, he now questioned the possibility of a consensus on past reality. As he wrote to John Adams in 1786: "The same facts impress us differently. This is enough to make me suspect an error in my process of reasoning tho' I am not able to detect it." For the skeptical Jefferson of the late eighteenth century, historical factuality was a difficult matter that relied on limited human "processes of reasoning," which were potentially dissimilar in different individuals. Not until a moment of consolidated historical certainties in the nineteenth century would Jefferson tell the same correspondent, "About facts, you and I cannot differ; because truth is our mutual guide."[33]

Historical Skepticism

Especially during the crises of his middle and later years, Jefferson evinced a skeptical attitude toward the literary narratives of philosophical history, sometimes with direct reference to the Pyrrhonian debates of earlier generations. Today, to be sure, he is mostly known for his antiskeptical stance, for his efforts, as he put it in an oft-quoted letter to John Adams, to "rid" himself of "useless" philosophical "Pyrrhonisms," such as through his reading of Thomas Reid, Dugald Stewart, or A. L. C. Destutt de Tracy.[34] However, one can only attempt to rid oneself of things one is aware of possessing. Concerning the problem of historical writing, Jefferson appears to have been burdened by a skepticism that was especially hard to subdue. It often propelled him to look for an alternative to the kind of historiography he had admired in his youth. In some cases, his skeptical emphasis on the limits of philosophical history would be expressed with an even-tempered openness to historical difference. In others, his doubts would verge on anger and even desperation. Two letters from his time in France that Jefferson wrote with some emotional investment but eventually failed to send exemplify two shades of his historical skepticism, as well as two different attempts to move beyond the confines of a didactic historiography.

A relatively serene instance of a skeptical mood can be found in the context of an essay on English prosody composed during Jefferson's stay in Paris and addressed to his friend from the Revolutionary War, the Marquis de Chastellux. There is evidence that Jefferson eventually failed to send to the author of *De la félicité publique* both the introductory letter from October 1786 and the draft of his *Thoughts on English Prosody*. The condition of the manuscript suggests, however, that he invested a considerable amount of time and reflection in the composition of the essay.[35] As he explained to Chastellux, he had begun his reflections on English prosody in order to defend a position he discovered to be untenable in the process of composing his *Thoughts:* "I began with the design of converting you to my opinion that the arrangement of long and short syllables into regular feet constituted the harmony of English verse: I ended by discovering that you were right in denying that proposition. The next object was to find out the real circumstance which gives harmony to English poetry and laws to those who make it." In this case, Jefferson did not permit himself to be frustrated by the experience of seeing his rhetorical "design" thwarted and admitting that a Frenchman had prevented him from "an error in my native tongue."[36] His mistake, as he claimed in a cosmopolitan spirit, only made him more curious to find out the "real" factor giving structure to English verse.

This factor, he realized, involved an intricate historical problem. Jefferson accordingly introduced his *Thoughts on English Prosody* on a skeptical note, with the somewhat sententious comment that "error is the stuff of which the web of life is woven: and he who lives longest and wisest is only able to weave out the more of it," and by recounting his recent experience that "probability like other human calculations, has been set aside by events." As introductory phrases, such references to the fallibility of human reason may be read as conventional expressions of modesty. Yet especially Jefferson's emphasis on historical contingency–his claim that unpredictable "events" could disprove "human calculations" such as probability–revealed a recognition that his object of study might prove too complex for his personal training in philosophical history. As he claimed in the opening of his *Thoughts,* prosody was the result of a long development over time: "I suppose that the system of rules or exceptions which constitutes Greek and Latin prosody, as shown with us, was unknown to those nations, and that it has been invented by the moderns to whom those languages were foreign. I do not mean to affirm this, however, because you have not searched into the history of this art, nor am I at present in a situation which admits of that search."[37]

In historicizing, to some degree, the reception of poetry in antiquity as well as the methods of teaching ancient verse in modernity, Jefferson displayed, again, his insight into the "fundamentally different" conditions of ancient and modern cultures. The moderns, as he emphasized, had no direct access to antiquity, because its languages were "foreign" to them. In this context, Jefferson went beyond the historicization of particular political institutions such as that of the temporary dictatorship in *Notes on Virginia.* His appreciation of historical difference also involved complex mental operations in the past and present: the modern methods of prosody and the reading or hearing processes involved in the ancient reception of poetry. This emphasis on historically variable perception processes in the "foreign" countries of the past went to the roots of historical difference and left behind, to some degree, the tautology of a universal human nature that had been fundamental to the moral function of philosophical history.

Jefferson was aware that "the history of this art" would require a lot of time and an intensive "search," a search that would be incompatible, if taken seriously, with his office as ambassador. By implication, the historical study of prosody was no longer the part-time work of a dilettante, but required the full time of an expert. Jefferson's doubts about his friend's and his own capabilities differed sharply from his earlier statements on the simple imitation processes involved in the study of philosophical history. His critical interest in the history of languages–in ancient European and Native American languages, as well as Anglo-Saxon and seventeenth-century English[38]–was part of a tradition ranging from the antiquarian philology of the previous centuries to, eventually, the professionalization of history in the nineteenth century. Jefferson's interpretation of linguistic change as a key to the differences of the past promised insights into the intricacy of historical processes that were not the less profound for consciously avoiding claims to an ultimate moral or national "truth."

In this letter, Jefferson's skepticism implied a potential alternative *within* the genre of historiography, pointing to a study of history that no longer took for granted, as did his early philosophical history, the primacy of its didactic function. Moving beyond a narrow moral and national context, Jefferson freely admitted to being uncertain about his historical outlook. It may ultimately have been this uncertainty that kept him from sending the text to his French correspondent. On the whole, however, his doubts seem less to have distressed him personally than to have made him eager for further research. As documented by another unsent letter from this period, by contrast, there were moments when Jefferson's historical skepticism could, by his own account, make him feel "agonies." Half-anonymous as this letter to the editor of the *Journal de Paris* already was–it was signed "An American"–it seems never to have reached its addressee, as Jefferson probably considered it imprudent to have it published. As the numerous deletions, corrections, and interlineations in the manuscript indicate, he apparently wrote it in a moment of fury occasioned by the *Journal*'s publication, on the same day, of an anonymous review of a historical work called *Les Ligues Achéene, Suisse et Hollandoise, et Révolution des*

Etats-Unis de l'Amérique comparés ensemble, by a certain Mayer. Jefferson was so enraged by the summary of this comparative historical study that he no longer expressed hope for an alternative within historiography, but thought of abandoning historical reading altogether:

> When young, I was passionately fond of reading history and travels. Since the commencement of the late revolution which separated us from Great Britain, our country too has been thought worthy to employ the pens of historians and travellers. I cannot paint to you, Sir, the agonies which these have cost me, in obliging me to renounce these favorite branches of reading and in discovering to me at length that my whole life has been employed in nourishing my mind with fables and falsehoods. For thus I reason. If the histories of d'Auberteuil and of Longchamps . . . can be published in the face of the world, and can be read and believed by those who are cotemporary with the events they pretend to relate, how may we expect future ages shall be better informed? Will those rise from their graves to bear witness to the truth, who would not, while living, lift their voices against falsehood? If cotemporary histories are thus false, what will future compilations be. And what are all those of preceding times?[39]

Personally connected as Jefferson saw himself to the space, time, and action of the events that historical narratives like Mayer's work "pretended to relate," he was shocked by the huge gap that seemed to open between "history" referring to the genre of historical writing (*historia rerum gestarum*) and "history" referring to the historical events (the *res gestae*) themselves. The part of the review that triggered his historical disillusionment, presented by Jefferson in the terms almost of a disappointed love or family relationship, is very interesting. The reviewer had claimed that Mayer's *Histoire* fulfilled the Aristotelian function of historiography, presenting particular facts of the American Revolution ("des particularités qui sont peu connues").[40] He mentioned one of these little-known historical particulars: the individual who had earned eternal fame by making possible the passage of the Declaration of Independence, according to Mayer (and uncontested by the reviewer), "fut John Dickinson. L'Amerique lui doit une reconnoissance eternal [*sic*]; c'est Dickinson qui l'a affranchie."[41] The historical detail that so enraged Jefferson that he concluded the letter, along with his farewell to the editor of the *Journal de Paris,* with his "Adieus to History," concerned the lasting reputation of a historical personality whom America, according to Mayer, owed eternal gratitude.

Before the party struggles of the 1790s, to be sure, Jefferson's own connection to the Declaration of Independence had not yet become a "fact" of American history. In 1787 Jefferson was not generally considered to be the "author" of the document (as he would claim to be in his *Epitaph*),[42] nor was he consistently identified as its "penman" or at least as a member of the drafting committee. In the histories of the American Revolution from the 1780s, even those written by personal correspondents such as William Gordon, David Ramsay, François Soulés, and Philip Mazzei, Jefferson remained a relatively marginal figure.[43] He had modestly refrained from advising these historians to foreground his own role in drafting the Declaration, just as, for that matter, he had refrained from giving the editor of the *Journal de Paris* more information on the individuals constituting the "Committee to prepare a declaration of independance."[44] But to have another man immortalized as the father of the Declaration of Independence–especially a moderate like Dickinson, who was popular among the French audience of the *Journal* but who had, after all, refused to sign the document[45]–was too much even for Jefferson, with his studied republican self-effacement.

His skepticism toward philosophical history, it may be inferred, was also part of a difficult process of becoming concerned about the future history of the American Revolution in general and of his own place in it in particular. What the self-identified "American" of the letter to the *Journal de Paris* criticized as historical misrepresentations to a European audience was problematic from a personal as well as a national point of view. Contrary to the cosmopolitan fallibilism evinced in Jefferson's *Thoughts on English Prosody,* the review of Mayer's *Histoire* shook his faith in historical writing generally. If historical works with blatant factual errors could remain unquestioned even by a contemporary reading public, histories referring to a past reality without living witnesses could hardly be expected to be more reliable. As Jefferson put it in the letter to the *Journal,* "Will those rise from their graves to bear witness to the truth, who would not, while living, lift their voices against falsehood?" In the case of Mayer's *Histoire,* ironically, it turned out that Jefferson lifted his own "voice against falsehood" mainly from beyond the grave, by preserving the unsent letter among his papers. The incident appears to have intensified his search for an alternative to philosophical history, but this time outside the medium of historical writing: his disappointments with historiography during his stay in Paris were accompanied by an increased interest in the visual arts as a medium of historical representation. On the day after he wrote his impassioned letter to the *Journal,* Jefferson wrote to the young painter John Trumbull urging him to leave London at once and visit him again in Paris as soon as possible: "You have only to get into the Diligence and in 4. days you are here."[46]

On this occasion, Jefferson used the Salon of 1787 as an incentive for Trumbull to come to Paris immediately. As their frequent correspondence from that time demonstrates, however, both men were already engaged in a dialogue about the visual representation of American history.[47] Originally, Trumbull's plans for a series of paintings depicting the American Revolution had merely included battle scenes. It was probably only after Jefferson's suggestion of the topic in the preceding year that he had begun preparations for the later-famous painting *The Declaration of Independence* (1786–1820)–under Jefferson's patronage, in residence at Jefferson's Hôtel de Langeac, and according to Jefferson's "information and advice," particularly by the aid of a floor plan drawn by Jefferson. Trumbull finally accepted the French minister's repeated invitations the following winter, when he painted his portrait for the *Declaration.*[48] The result of this patronage of the young painter was a work that presents Jefferson as the central member of the drafting committee. His tall figure is the focal point of Trumbull's arrangement of the Continental Congress, distinguished not only by his central position and the color of his hair and vest but also by the fact that he is holding the draft of the Declaration in his hands, in the act of presenting it to John Hancock. Jefferson's prominent place in the painting thus contrasted sharply with his obscure connection to the Declaration of Independence in the written histories of the 1780s. His show of modesty on one of the rare occasions when he was personally connected to the Declaration in these years–a "Fourth of July Tribute to Jefferson," probably drafted by Joel Barlow for a private celebration at Jefferson's hôtel in 1789[49]–appears rather incongruous with his conspicuous role in Trumbull's painting: in reply to praise for having "added a peculiar lustre to that declaratory act which announced to the world the existence of an empire," Jefferson humbly observed that his own little contribution was "best in the shade. The light of the picture is justly left to others."[50]

Thus, Jefferson's symbolic share of enlightenment came to depend, for a while, on the medium of historical representation. The painted history of Trumbull's *Declaration of Independence,* still in miniature in the 1780s, anticipated Jefferson's conspicuous position as the document's "author," a historical interpretation that moved to the foreground in the 1790s and eventually became institutionalized, as it were, by Trumbull's monumental version of the *Declaration* in the Capitol Rotunda. By the end of his life, Jefferson's historical reputation had materialized in painting, sculpture, and architecture, as well as writing. From the vantage point of the 1780s, however, he was confronted with a precarious discrepancy between his place in written and in painted history. This discrepancy was also the result of a theoretical problem. When eighteenth-century critics from Diderot to Lessing interrogated the Horatian formula *Ut pictura poesis,*[51] their arguments against the basic similarity of the "sister arts" of poetry and painting (including the historical sisters of a literary historiography and historical portraiture)[52] emphasized the factor of time as the crucial element distinguishing narrative works from the visual arts. Unlike literary narratives, modern paintings did not depict a "succession of incidents," but were "confined to a single instant in time," as Lord Kames put it.[53] This stress on the temporal element implied what may be described as a neo-Aristotelian distinction between mimetic works and a nonmimetic historiography: history paintings could legitimately conflate events in time and space in their mimetic depiction of the probable, while historical writing had to focus on the representation of historical events in their accidental succession in time.[54]

Authors and theorizers of philosophical history, as has been discussed, had tried to overcome Aristotle's restrictive concept of historical writing by imposing their own mimetic order on the events, making vague conjectures about the past and giving a paradoxical priority to the future. Jefferson had praised these literary methods in his early years, but when his own actions were becoming subject matter for historians in the 1780s, he condemned such generalizations as "fables and falsehoods," as in his letter to the *Journal de Paris.* Meanwhile, he continued to tolerate them in historical painting until the end of his life. For instance, the aged Jefferson was confronted by a grandson of Samuel Adams, Samuel Adams Wells, who expressed, in the course of their exchange on revolutionary American history, his strong disapproval of the factual inaccuracies in Trumbull's *Declaration of Independence.* In his reply, Jefferson defended Trumbull's Capitol version on the grounds of *licentia pictoris:* "The painting lately executed by Col. Trumbull I have never seen: but as far back as the days of Horace at least we are told that 'pictoribus atque poetis, quidlibet audendi semper fuit aequa potestas.' he has exercised this licentia pictoris, in like manner, in the surrender of York, where he has placed Ld. Cornwallis at the head of the surrender, altho' it is well known that he was excused by General Washington from appearing."[55]

As this explicit defense of Trumbull's mimetic art suggests, Jefferson's expectations on the two historical media diverged significantly (much in contrast, apparently, to those of his correspondent). Consequently, Jefferson could imagine historical painting as the better alternative to historical writing, as a way to represent the probable and the morally instructive, while avoiding the disappointing "falsehoods" in contemporary historiography. Paintings presented only a single "instant" and made no problematic claims to foresight, carrying along the ignorant conjectures, falsifying comparisons, and detail-defying analogies that Jefferson had come to despise in works of philosophical history such as Mayer's *Histoire.* As openly mimetic works that did not pretend to relate what had actually happened, history paintings could not really "lie." Moreover, according to Jefferson's Kamesian aesthetics, painting generally appeared as a medium that was more likely to achieve advantageous didactic effects: it was better equipped to make essential historical events accessible, through the sense of sight, to the moral sense.[56]

Jefferson's claim that he had never seen Trumbull's *Declaration* still reveals, to be sure, a general uneasiness with representations of his own role in the American Revolution. While this claim to ignorance was justified, in a narrow sense, with respect to the painting's recent version in the Capitol, it deliberately obscured information that must have been known to Jefferson: not only had he once drawn a floor plan for the painting but there was, after all, an engraving of Trumbull's work displayed in his own Monticello (figure 4). Jefferson's not uncharacteristic mode of indirection in this instance–accompanied by a recourse to the murky realm of half-truths–suggests an additional reason why he regarded philosophical history as a flawed mode of historical representation. This reason concerned the depiction of republican virtue in the construction of a national history of American progress. By patronizing a nonlinguistic work such as Trumbull's *Declaration,* Jefferson could convey his republican merit with relative clarity. Achieving the same effect in language was more difficult because, as the twisted self-denial of Jefferson's letter to the *Journal de Paris* suggests, republican virtue threatened to turn into its opposite in the moment of its linguistic representation. As discussed in the previous chapter, readers of philosophical history were expected to be "bewitched" by master writers who conveyed historical events by the illusionistic "magic" of their style, a style that was likely to misrepresent historical facts, whether motivated by the writer's own political bias or by his having been "bewitched," himself, by previous historians. It was extremely difficult to interrupt this vicious circle, even when contemporary readers such as Jefferson had been "on the

Figure 4. *The Declaration of Independence of the United States of America, July 4th, 1776,* after John Trumbull, by Asher B. Durand, engraver, 1823. (Monticello/Thomas Jefferson Foundation, Inc. Photograph by Edward Owen)

spot."[57] Such personal witnesses could find it impossible to "lift their voices against falsehood," because their own actions were likely to be concerned. Lest they appear guilty of an egotistical ambition that could seriously compromise the reputation of a more or less self-denying virtue,[58] men like Jefferson tended to silence themselves, hoping that their historical achievements would be established as historical facts by others.

As a result, Jefferson often worried about whether he would gain the place in American history that he thought he deserved. At the time of the letter to the *Journal de Paris,* of course, he did not know that he would eventually be known in schoolbooks as the author of the Declaration of Independence or as the third American president. It seemed likely, at that point, that his drafting of the Declaration would at best become a footnote in the historical record–or that John Dickinson would eventually achieve more lasting historical fame than he. And if Jefferson's republican modesty had prevented him from influencing the historical record on his own account, it could be inferred that virtuous men in the past had likewise failed to achieve lasting fame. Many of the historical narratives Jefferson had admired in his youth, it seemed possible, had merely depicted the egotistical villains who had forced their way into historical memory–and the John Dickinsons, who had survived in the record by accident, due to the many mistakes historians were apparently prone to make.

Historiography, even a didactic historiography purporting to produce "better men, and better citizens,"[59] appeared to have an unfortunate tendency to misrepresent, in particular, the best citizens in history–self-denying republicans like Jefferson who had been virtuous enough to subordinate themselves (and their place in history) to the greater good.[60] In order to promote their historical reputation, therefore, it often seemed safer to find the means outside of philosophical history. For Jefferson, it was best possible by indirection: by telling the editor of the *Journal de Paris* to look up the "Journals of Congress" in a letter that never reached him, by postponing the publication of his historical notes until after his death, and by giving Trumbull hints for a history painting. Instead of meddling in the historical discussions of the present–the less successful strategy of his more hot-tempered friend John Adams–Jefferson thus began to direct his efforts at the future, creating potential source material for future historians. In this spirit, he tended to suppress many of his statements and withhold their publication until he could expect them to acquire the impersonal authority of a "voice" lifted from beyond the grave, as in the letter to the *Journal de Paris.* However, Jefferson realized that this alternative to philosophical history had its own perils. Original artifacts could easily be damaged or destroyed. And unless they were printed, original manuscripts always threatened to be "lost" entirely, subjected as they were to the "daily havoc" and "the waste of time."[61]

Fragments, Ruins, and Remains

"On the subject of the history of the American revolution, you ask Who shall write it? Who can write it? And who ever will be able to write it? Nobody; except merely it's external facts. All it's councils, designs and discussions, having been conducted by Congress with closed doors, and no member, as far as I know, having even made notes of them, these, which are the life and soul of history must for ever be unknown."[62] In this 1815 reply to John Adams, Jefferson concurred with his correspondent that the narratives of a literary historiography–in this case, Carlo Giuseppe Guglielmo Botta's *Storia della guerra americana* (1809)–were uncertain depositories of historical knowledge. If history was to "survive" in the present and future, it would not be in the lifelike illusions created by the literary techniques of philosophical history, such as Botta's recourse to the ancient tradition of inventing "historical" speeches. History could "live" only in the occasional notes, letters, and memoranda written by the historical actors themselves. If these eyewitness accounts were missing, Jefferson claimed, the "life" and the "soul" of history would vanish. The past, in other words, would be literally "dead" and "lost."

Given his actual efforts to shape the future memory of the revolutionary years (even without the benefit of notes from the revolutionary Congress), it is questionable whether Jefferson was fully convinced that "nobody" could ever write anything but an external record of the American Revolution. His emphasis on the importance of original documents, however, was probably sincere, as it was part of a lifelong interest in archival preservation. In his early years, this interest had largely been restricted to documents pertaining to the colonial history of Virginia, at first falling into his hands accidentally and then collected more systematically.[63] As Jefferson later put it, his concern about the "loss" of colonial records had begun "early in the course of my researches into the laws of Virginia," when he "observed that many of them were already lost, and many more on the point of being lost." By 1796, as in a letter to George Wythe, he advocated the preservation of the "remains" by a systematic "multiplication of printed copies" in order to avoid the "loss in these precious monuments of our property, and our history" and to prevent their "future loss."[64]

This obsession with a threatening "loss" of historical knowledge was, to some extent, a modernized version of an earlier approach to history, that of the antiquarian historians of the previous generations whose "learned lumber"[65] had appeared so ridiculous to Jefferson as a young man. With the experience that the past was a "foreign" country with "fundamentally different" historical conditions whose unique quality threatened to disappear in the future, Jefferson tended more often to acknowledge the importance of antiquarian research. His preservational impulse intensified during the late eighteenth century, when his personal disappointments with philosophical history suggested an additional function of preserving "remains" from the past. Regarding recent American history, especially his own involvement in it, a conscientious mode of collecting documents could be used as a cautious alternative to a personal engagement in the historiographical debates of the day.

This alternative became important to him in the 1780s and 1790s. Significantly, Jefferson tended to abstain from composing larger historical or autobiographical narratives during these decades, although he retired twice from political office, from his governorship of Virginia in 1781 and from the office of secretary of state in 1794. In both cases, he probably felt some need for personal vindication, considered his leave from political life definite, and developed a penchant for announcing his own approaching death. In neither of these retirements, however, did he consistently employ what seemed like the hindsight perspective of his political life for the composition of his memoirs.[66] Instead of working on a personal narrative, he directed his efforts at collecting and preserving documents of his experiences, hoping to create the source material for future historians. In 1779 and more comprehensively in 1783, he began to keep records of both his incoming and his outgoing letters. Aided by James Watt's copy press and the polygraph, he preserved a large amount of his voluminous correspondence for the use of future historians.[67] Especially during the tumultuous 1790s, moreover, Jefferson kept short "memoranda" of political conversations–later known as the *Anas*–as yet failing to provide a didactic narrative connecting these fragments.

The effort to unify these short pieces within a coherent interpretation of his life would not come until two decades later, in his 1818 explanations to the *Anas* combined with his 1821 *Autobiography.* Looking back to the 1790s, the aged Jefferson described shifting priorities in the documentation of his life:

> In the earlier part of my acting in that office [of secretary of state] I took no other note of the passing transactions: but, after awhile, I saw the importance of doing it, in aid of my memory. Very often therefore I made memorandums on loose scraps of paper, taken out of my pocket in the moment and laid to be copied fair at leisure, which however they hardly ever were. These scraps therefore, ragged, rubbed, & scribbled as they were, I had bound with the others by a binder who came into my cabinet, did it under my eye, and without the opportunity of reading a single paper. At this day, after the lapse of 25 years, or more, from their dates, I have given the whole a calm revisal, when the passions of the time are past away, and the reasons of the transactions act alone on the judgement. Some of the informations I had recorded are now cut out from the rest, because I have seen that they were incorrect, or doubtful, or merely personal or private, with which we have nothing to do.[68]

According to this retrospective narrative, Jefferson's approach to historical preservation underwent three phases. In his earlier political life, he merely collected official documents on occasion, without taking any "other note of the passing transactions." After some time in this practice, he realized the need to also write "memorandums" of political meetings, conversations, and so on, for the sake of his "memory" (with the possible double sense not only of his own faculty of memory but also of the way he would be remembered by future generations). By 1818 he had again changed his attitude, freely admitting that he had decided to make selections from among the original memoranda according to his own improved "judgement."

While this open-hearted acknowledgment of having interfered with a supposedly authentic documentation is itself interesting for Jefferson's more literary emphasis in his final retirement (as discussed in chapter 6), it is at first necessary to clarify how precisely he expected his memoranda to prevent the "loss" of the past. His short pieces were "the life and soul of history," according to the definition in his letter to Adams, because they were the only way to preserve the knowledge of political events that had passed "with closed doors" and involved–also in republican America, as Jefferson realized–only small groups of people. In writing these notes for future historians, his priorities shifted between two major senses of the term *history:* from history understood as historical writing to history referring to the historical events themselves. His memoranda were not supposed to influence, by their "artful" literary composition, the future behavior of their readers. Instead, Jefferson foregrounded specific characteristics of these texts that distinguished them from the rhetorical function of philosophical history. He emphasized that in contrast to the polished "design" of a universalizing grand narrative, his notes were contingent remains from a past reality, consisting in nothing but "loose scraps of paper, taken out of my pocket in the moment." Without the elegant trompe l'oeil of philosophical history, the fragments were part of the reality of a unique moment in the past. By drawing attention to the unevenness of their tangible materiality (described, in almost onomatopoetic expressions, as "ragged, rubbed, & scribbled"), Jefferson created a quasi-Romantic aesthetic of authenticity.[69]

Presenting his memoranda as fragmentary, unmediated remains, he tried to divest these pieces of his own writing of any impressions of personal intentionality. His uncommonly detailed information on the binding process served a similar function: since they had been bound, as he tried to demonstrate, nothing could have been added or removed (at least not until his own activities in 1818). Jefferson's seemingly paranoid insistence that even the binder had not had a chance to read any of the memoranda implied that future readers would gain access to previously undisclosed texts in their original, "undesigned" state. Nineteenth-century historians were to consider themselves distinguished as first readers of what Jefferson had supposedly written in a hurried moment in the 1790s. Posthumously, an eyewitness to and a key actor in their early national history would offer them a share in secret information on their history. Through his comments on the binder's ignorance of the memoranda he thus sought to establish a direct, personal link between the moment in which the memoranda had been committed to paper in the late eighteenth century and the moment in which they would be read by later generations.

Jefferson's efforts thus to prevent the "loss" of the past did not concern his papers alone. Along with written documents, he began to be occupied with visual artifacts as possible sources of historical knowledge in the 1780s. While in Paris and London, he met and corresponded with painters, sculptors, architects, and antiquarians such as John Trumbull, Benjamin West, John Singleton Copley, Mather Brown, Richard and Maria Cosway, Jean-Antoine Houdon, Jacques-Louis David, Elisabeth Vigée-Lebrun, Hubert Robert, Charles-Louis Clérisseau, Josiah Wedgwood, and Pierre-François Hugues d'Hancarville. He visited studios and galleries and attended exhibitions, began to buy paintings and prints (also "of the historical kind"),[70] and commissioned portraits and busts.[71] Jefferson regarded paintings like Trumbull's *Declaration of Independence,* as has been said, as mimetic works that could legitimately conflate events in space and time in their imitation of the probable. In another respect, however, such works might also be expected to be considered "authentic" historical documents later on. By keeping his correspondence with Trumbull among his papers, Jefferson enabled later readers to know about his personal friendship with the painter. In hindsight, Trumbull would be regarded not only as a contemporary artist but also as an eyewitness who had personally seen, if not the event itself, at least many of its protagonists.

For Jefferson, this firsthand preservation of personal traits was an additional advantage of the visual arts. It afforded the best means of transmitting the knowledge of physical details through space and time. Such information could easily be lost, as Jefferson knew had happened in the case of many ancient writers whom he admired. When the events of the American Revolution were becoming subject matter for historians, accordingly, he began to commission original paintings and copies of existing paintings of men whom he considered to be important for American history, such as Adams, Paine, and the "three greatest men that have ever lived"–Bacon, Locke, and Newton.[72] It was in this period that Jefferson also sat for his own first portrait, by Mather Brown, which was quickly followed by such likenesses as a marble bust by Houdon and several miniatures by Trumbull, including three gift copies, for Maria Cosway, Angelica Church, and his daughter Martha.[73]

Tied to the themes of vanity and mortality, separation and loss, these personal gift copies exemplified the sentimental aspect of eighteenth-century conventions of portraiture. Like Jefferson's scribbled "scraps of paper," they sought to prevent the loss of the past by retaining a material link to an absent person or to a past moment in individual or collective history. This connecting function of visual remains was also crucial, on both a national and a personal level, in another domain of the fine arts that often inspired Jefferson to reflect on the passing of time: architecture. On the national level, he was instrumental in modeling the Virginia capitol in Richmond on a ruin from the Roman Empire. As he wrote to Madison in 1785, he considered the ancient Maison Carrée, in Nîmes, an appropriate model because it was best equipped to meet the aesthetic judgment of future generations. Parallel to his eighteenth-century perspective on the Roman temple, he anticipated the perspective of later centuries, when the Virginia capitol, possibly then a ruin itself, would be regarded as a founding monument of American civilization. As such, it had to be a "proof of national good taste" rather than "a monument of barbarism."[74] By his choice of an ancient model for a modern American building, then, Jefferson was again trying to create sources for future historians on the important question of "national" taste.[75] Architecture–whether of the Virginia capitol, the University of Virginia, or his own Monticello–was for him another means, apart from his papers and his portrait collection, to try to influence the historical reputation of his period (and himself) in the eyes of posterity.

Yet architecture did not consistently appear to promise continuity and permanence to Jefferson. In the spring of 1787, when he traveled through Provence, he was prepared to view the scenes of Clérisseau's *Antiquités de la France: Première partie. Monumens de Nismes,* which he had recently bought. Long before traveling to Nîmes, "gazing whole hours at the Maison quarrée, like a lover at his mistress,"[76] he had been convinced of the timeless aesthetic significance, the classical simplicity and nobility,[77] of ancient Roman ruins in general and the Maison Carrée in particular. What was new in his three-dimensional experience of these buildings, "on the spot" in southern France, was the physical details of their decay. As documented in his letters and travel notes, Jefferson "immersed"[78] himself in the materiality of the ancient remains, taking elaborate measurements of the ruins and analyzing the quality of the bricks. He did so not only to transmit this information in space–from France to Virginia–but also to preserve precious knowledge in time–from antiquity to posterity. As in his scribbled "scraps of paper," it was the material details of the historical remains whose ambivalence likewise fascinated and frustrated Jefferson. While they qualified them as authentic parts of a past reality, they also revealed the fragility of these buildings, foreshadowing their possible loss. In a letter to Madame de Tessé usually noted for its humorous expressions of admiration for the Maison Carrée, Jefferson also referred to his anxieties:

> I am glad you were not there; for you would have seen me more angry than I hope you will ever see me. The Pretorian palace, . . . totally defaced by the Barbarians who have converted it to it's present purpose, it's beautiful, fluted, Corinthian columns cut out in part to make space for Gothic windows, and hewed down in the residue to the plane of the building, was enough, you must admit, to disturb my composure. . . . Would you believe, Madam, that in this 18th century, in France, under the reign of Louis XVI, they are this moment pulling down the circular wall of this superb remain [the arena at Orange] to pave a road?[79]

It was not often that Jefferson allowed himself to express personal vexation in his letters. The gratification of his sentimental "immersion" in antiquity was accompanied not only by his characteristic disdain for a "Gothic" age but also by intense apprehensions about the savagery of "this 18th century." His fear of "Gothic" barbarism pervades the letter and returns, in its final paragraph, on the imaginary level of a time-travel fantasy, which may have been inspired by a passage in Gibbon's *History of the Decline and Fall of the Roman Empire* (1776-88).[80] Jefferson overstated his sentimental preoccupation with ancient remains by claiming to be "immersed" not only spatially in "antiquities" but also temporally in antiquity itself. This immersion allowed him this strangely anachronistic perspective:

> Were I to attempt to give you news, I should tell you stories one thousand [*sic*] years old. I should detail to you the intrigues of the Caesars, how they affect us here, the oppressions of their Praetors, Praefects, &c. I am immersed in antiquities from morning to night. For me, the city of Rome is actually existing in all the splendor of its empire. I am filled with alarms for the event of the irruptions daily making on us, by the Goths, the Visigoths, Ostrogoths, and Vandals, lest they should re-conquer us to our original barbarism.[81]

FIGURE 5. *La Maison carrée, les arènes, et la tour Magne à Nîmes,* by Hubert Robert, 1787. (Musée du Louvre, Paris. Inv. 7468. Photograph by Gérard Blot, RMN)

In his ironical time travel Jefferson tried to overcome the loss of the past by constructing a condensed imaginary moment that placed him simultaneously at the height of the Roman Empire, at the moment of its destruction, and even in the year 787. This conflation of time levels was reminiscent of contemporary paintings that he admired in Paris, as by Hubert Robert, who was able to depict the Maison Carrée as a modern ruin illogically peopled by figures in ancient dress (figure 5). It was taken up by Madame de Tessé in her response, in which she even surpassed Jefferson's playful incongruities by claiming that she had been inspired to travel through the centuries, from ancient Egypt to the remote future, until she had found herself "en Société avec [H]omère, Licurgue et Solon."[82]

In all of these cases, the identification with antiquity functioned, as in the opening of *Les ruines* (1791), by Jefferson's correspondent Constantin François de Chasseboeuf Volney, only by a breakdown of rational time perception. As Jefferson put it in a contemporaneous letter to Madame de

Tessé's protégée Madame de Tott, his glimpse into the ancient world occasioned by Drouais's painting *Marius prisonnier à Minturnes* had "fixed" him "like a statue a quarter of an hour, or half an hour, I do not know which, for I lost all ideas of time."[83] In his rapture, Jefferson had supposedly become equivalent, himself, to a work of art transcending the linear order of time. Such an ironical overstatement was necessary in his identification with Roman antiquity because in serious contexts his views on its political history–especially, of course, of the Roman Empire–were characterized by a conviction of its inferiority. The history of the Roman Empire was for Jefferson, as for many of his contemporaries, culturally ambivalent (a history of "grandeur et décadence," in Montesquieu's terms) and politically destructive (a history of "decline and fall," in Gibbon's terms). In the letter to Madame de Tessé, Jefferson thus referred to the (cultural and aesthetic) Roman "splendor," on the one hand, as well as to the (political and moral) "intrigues," "oppressions," and "alarms," on the other. To his mind, Roman history was always defined by its outcome, invariably destined to end by a "re-conquering" to (aesthetic *and* political) "original barbarism." Interestingly, Jefferson's letter to Madame de Tessé concluded with a link between late Roman antiquity and French modernity: "Gothic barbarism" threatened to destroy his beloved classical architecture as much in his own period as in his imaginary antiquity.[84]

Jefferson's analogy between the late Roman Empire and the final years of the *ancien régime* may not appear far-fetched in retrospect. His anxieties concerning "the Goths, the Visigoths, Ostrogoths, and Vandals" even seem rather prophetic, considering an attitude toward the past, and toward remains from the past, that would become excessive in France during the first half of the following decade. Although no doubt largely due to the hyperbole of the letter to Madame de Tessé, it is remarkable that Jefferson referred to the "Vandals" in particular to mark the destruction of precious artifacts from the past. When during the French Terror buildings, tombs, and statues fell prey to outbreaks of revolutionary iconoclasm, Jefferson's acquaintance Henri Grégoire coined the term *vandalisme* in order to condemn this blind destruction. (As Grégoire would later claim, "Je créai le mot pour tuer la chose.")[85] This "vandalism," it can be argued, was closely linked to the modern impulse to preserve, restore, and historicize material links to the past. In a sense, the iconoclasm of the 1790s was also a continuation of the enlightened fascination with ancient ruins: Jacobin "vandals" produced the ruins that already existed from antiquity, thereby creating the material necessary for a historicization of other periods as well.[86]

In a way that Jefferson would probably have found confusing, the Jacobin "vandalism" lamented by Grégoire in 1794 was often directed at medieval, indeed "Gothic," architecture and sculpture, such as in churches and monasteries. What united Jefferson's and Grégoire's attacks on modern vandalism, however, was their indignation about an uncivilized lack of historical consciousness that resulted in a mindless destruction of everything that did not serve the needs of the present. Both men thought their own period superior to the past, but both men also sought to preserve its remains: Jefferson looked down on the ethics and politics of the Roman Empire yet deeply regretted the destruction of its ruins; Grégoire opposed feudal privileges yet publicly condemned the destruction of medieval artworks. In their criticism of modern vandalism, Jefferson and Grégoire thus began to separate the past from its didactic function for the future. To some extent, they came to cherish its material remains for their own sake, simply for their relationship to a unique moment in time. The enlightened disdain for a less enlightened past could thus be accompanied, even for progressives like Jefferson and Grégoire, by growing fears of losing remaining links to this past–whether brought about by economically motivated reform, as by the road pavers Jefferson criticized in Orange, or, in radicalized ways, by the ideologically motivated demolitions during the French Terror. In each case, the apprehension of an impending loss resulted in an intensified urge to "preserve and restore," as Jefferson emphasized in the letter to Madame de Tessé.

Jefferson's preservational impulse may have been least successful with regard to the history of American "Indians."[87] In 1809 he deplored an "irreparable misfortune" that was, to some extent, paradigmatic of his views on American "antiquity." Over the course of many years, Jefferson had collected Indian vocabularies, which were almost entirely destroyed by a thief who threw them into the James River shortly before Jefferson had occasion to publish and thus preserve them for posterity. In terms similar to the tangible authenticity of his scribbled "scraps of paper" in the *Anas,* Jefferson described the tragedy of these unique historical documents that had flowed down the river of time: "Some leaves floated ashore & were found in the mud; but these were very few, & so defaced by the mud & water that no general use can ever be made of them." In the "wrecks of my loss," as he put it, only "fragments" remained.[88] A quarter-century earlier, in *Notes on the State of Virginia,* Jefferson had already described an American antiquity that was more definitely absent from the present than a European antiquity whose ruins still partook in the fashionable world of the Enlightenment.

Jefferson invariably linked the topic of precolonial American remains to the theme of loss: the certain loss of the Indian societies themselves (as he saw their only future in their assimilation to Anglo-American culture) and the probable loss of any precise information about them (unless more historical "search" and documentation was done quickly).[89] American antiquity, unlike its European counterpart, threatened to entirely "disappear in time."[90]

Although they still populated the larger part of the continent when he wrote his *Notes,* Jefferson chose to regard the Indians primarily as precious historical material in this context. He seemed to regret their disappearance mainly for the consequent loss of historical information: "It is to be lamented then, very much to be lamented, that we have suffered so many of the Indian tribes already to extinguish, without our having previously collected and deposited in the records of literature, the general rudiments at least of the languages they spoke."[91] As can be concluded from this (for today's sensibilities) rather cynical statement, Jefferson's interest in Indian languages was essentially an interest in "dead" languages, comparable to his interest in Anglo-Saxon, Latin, and ancient Greek. Query 11 is full of reflections on the passing of time, the antiquity of the supposedly "new world," and the long pedigree of the allegedly "new man." Jefferson's linguistic research convinced him of the "remarkable fact" that the great variety of radically different Indian languages could only have developed over a "long course of time" or even in "an immense course of time; perhaps not less than many people give to the age of the earth."[92] As he considered aboriginal American languages "of greater antiquity than those of Asia," he was inclined to believe that Asians were actually descended from the American Indians, rather than vice versa.[93]

In other contexts, even when he did not go as far as to explicitly reverse the relationship between the "old" and the "new" world, he emphasized the parallels between Indian culture and European antiquity. He identified Indian burial rites with those of the Greeks described by Homer or of the Persians described by Herodotus, their oratory with that of Demosthenes and Cicero, and the state of their arts and sciences with that of the Germanic tribes described by Tacitus.[94] According to Jefferson's Whiggish Anglo-Saxonism, he considered the freedom-loving tribes of Tacitus's *Germania* also to be the European ancestors of republican Americans.[95] On a metaphorical level, Jefferson could thus imagine the first Americans and Anglo-Americans to have descended from the same remote ancestors. This not only made it plausible for him to claim family relations with them in his

Indian addresses, to unite them with Anglo-Americans under the category "Man of America," and, in contrast to his views on American blacks, to tolerate their "racial" amalgamation.[96] It also allowed him to merge European and American history in ways that combined two advantages: His "new" world republic gained a historical dimension that went as far back as that of contemporary European nations. And fortunately, this ancestry was situated in "the most remote antiquity"–hence, unlike the immediate English ancestry, far enough in the past not to cause any political trouble in the present and future.

His Indian-Germanic ancestors seemed temporally so remote to Jefferson that he found few material traces of them in America, except for their graves: "I know of no such thing existing as an Indian monument: for I would not honour with that name arrow points, stone hatchets, stone pipes, and half-shapen images. Of labour on the large scale, I think there is no remain as respectable as would be a common ditch for the draining of lands: unless indeed it be the Barrows, of which many are to be found all over this country."[97] Jefferson's personal engagement in the excavation of these sole "Indian monuments," the mounds of their graveyards in Virginia, is documented by a long passage under query 11, which has been praised for its pioneer position in the invention of the stratigraphic method in modern archaeology.[98] However, there was also a more backward-oriented, almost morbid side to Jefferson's "immersion" in Indian graveyards and his close analysis of the fragmented skeletons. Jefferson used the terms *vestiges* and *remains* to refer to Indian artifacts, Indian languages, and even the diminished tribes themselves.[99] Through his investigations of remains in the second degree, that is, fragmented bodily remains of tribal remains, his historical interests moved one step further into the past. As he pointed out, the barrows had still played a role in Indian rites "about thirty years ago," but since the lands had become cultivated, the mounds were destroyed quickly, already being "much reduced in their height, and spread in width, by the plough." Due to the progress of Anglo-American *culture,* in the modern and the original sense of the term, even these last remaining links to Indian history–material remains containing bodily remains of tribal remains–already belonged chiefly to the past and would soon entirely "disappear in time."

The figure of the heroic yet inevitably vanishing Indian would, of course, play an important role in nineteenth- and twentieth-century American culture, following a centuries-old cult of the noble savage. In what was for some time the most controversial passage of the *Notes,* Jefferson introduced a noble savage into the text, providing a piece of "ancient" Indian oratory that could challenge "the whole orations of Demosthenes and Cicero."[100] The translation of a speech by "Logan, a Mingo chief, to Lord Dunmore" had appeared in American and European newspapers during the 1770s, receiving such renown that it had been used as "a school exercise for repetition," according to Jefferson.[101] In the version he quoted, Chief Logan–originally a "friend of the whites," who became their enemy after his family was murdered by frontiersmen–concluded his oration by turning himself into one of the last Indian "remains" Jefferson had described elsewhere in the *Notes:* "There runs not a drop of my blood in the veins of any living creature. This called on me for revenge. I have sought it: I have killed many: I have fully glutted my vengeance. For my country, I rejoice at the beams of peace. But do not harbour a thought that mine is the joy of fear. Logan never felt fear. He will not turn on his heel to save his life. Who is there to mourn for Logan?–Not one."[102]

Whether this extract was really part of a historical speech, or mainly a fabrication similar to other stage orations of the 1770s,[103] it fitted perfectly into Jefferson's image of "ancient" American history. Having taken his archaic revenge, Logan conveniently remained a good American patriot by delighting, after his defeat, at future peace for his "country." He voluntarily engaged in an effective self-historization, choosing to retreat completely into the past and connecting the physical extinction of his relatives with the necessity of his own death. It has been argued that, tragically, it was the Indians' adherence to the revolutionary maxim "Liberty or death" that contributed most to Jefferson's regarding them as his equals.[104] His Logan thus had no choice but to announce his own approaching death, a death that would not be mourned, since all his personal links to other "living creatures" had been destroyed. For the short time that was left to him, he was the last remain of a past that had been different in quality from the present and would be completely absent from the future.

Logan's moribund perspective seemed directly opposed to the didactic function of philosophical history, disinterested as he professed himself to be in his "felicity and future fame."[105] The Indian leader affirmed his position in a unique past, a past defined by rupture and difference, clearly separated from the rites of memory in a changed American society. In the very act of cutting the ties to the future by a memorable speech, however, Logan created an "Indian monument" that could later be memorized by Anglo-Americans like Jefferson as a heroic remain from their national history. So ingrained was this nationally constitutive "loss" of the past in Jefferson's thought that he began to speak of "remembrance" as early as

1781, even when addressing the Indians themselves. As might be said maliciously today, it was as if he already envisioned the visual remains of their culture displayed, long after its disappearance, in his entrance hall at Monticello. Jefferson, it seems, enjoyed "reading" American antiquity all the while serenely contemplating its loss. As he explained to the Indians, "I have carefully attended to the figures represented on the skins, and to their explanation, and shall always keep them hanging on the walls in remembrance of you and your nation."[106]

The Americanization of Philosophical History

In the late summer of 1789, shortly before leaving France to return to America, Jefferson wrote a text that abounds with reflections on time and history and is today regarded as one of his most complex political and philosophical arguments. In a long letter to James Madison, he explained the "self evident" principle " '*that the earth belongs in usufruct to the living;*' that the dead have neither powers nor rights over it."[107] This argument for generational sovereignty involved intricate legal, constitutional, financial and economic problems.[108] It stands out from Jefferson's other writings due to a claim of newness he rarely made regarding his own ideas: "The question Whether one generation of men has a right to bind another, seems never to have been started either on this or our side of the water."[109] Moreover, the argument on generational sovereignty can be understood as Jefferson's most systematic effort to reason out a gradualist conception of historical change–to develop a theory, as it were, for the two temporalities inherent in a universal "course" that contained "human events" and to achieve the much-needed balance between his universalist hopes for the future and his regard for the past. This balancing act between two "times," when it was successful, enabled him to assume particular rates of change for different groups of people without losing sight of his universal historical telos. However, Jefferson also experienced situations in which the poise of his construction was seriously destabilized. When the two temporalities appeared to diverge, he tended to express the uncanny sensation that time was accelerating. In the last decade of the century, Jefferson linked this acceleration to a heightened sense of urgency concerning, in particular, a solution to the slavery problem. The topic of slavery made it especially attractive for him to distribute the two temporalities contained by generational sovereignty in space and consolidate his idea of a "material" difference between the history of the American republic and the history of the rest of the world.[110]

This idea, I shall argue in this section, was at the root of Jefferson's "Americanization" of philosophical history in the nineteenth century.

As has been discussed, Jefferson often defined the relationship between past and present–most clearly in his imagined history of American "antiquity"–by rupture, loss, and death. Such sharp boundaries also characterized the generational thought in Jefferson's famous letter to Madison. Like Chief Logan, who linked his family's death to his own, Jefferson conceived of clear-cut "generations" by assuming simultaneous births and deaths for a group of contemporaries whom he imagined to "come on, and go off the stage at a fixed moment." Defined by their exact contemporaneousness "on the stage" of history, the members of each generation were for Jefferson equivalent to a "society" or "nation"–parallel again to Logan, who felt compelled to leave the stage after his tribal Indian nation had disappeared. Jefferson tamed the contingency of Logan's fate, however, by mathematically calculating the stage appearances of his political generations.[111] As a successful instance of his gradualism, he thus retained some rational control over the chaos of sentimental time. The universalizing grid he imposed on historical change enabled him to defend the claims of the present generation against the hereditary privileges of past generations (which was the pressing issue in France in 1789), as well as the claims of future generations against those of the present (as in questions of private and public debt in America). Thus, he could make a universal statement about the temporal boundaries between two generations that he held to be as significant–morally and legally, politically and financially–as the spatial borders between two different "societies" or "nations." As he put it in a universalizing present tense, "by the law of nature, one generation is to another as one independant nation to another."[112]

This political equation was also the theoretical basis for a difference in historical quality: each "temporal nation" had to take care of itself, satisfying its own needs and shaping its own institutions. As Jefferson emphasized, "No society can make a perpetual constitution, or even a perpetual law." Members of the same temporal nation could only authorize "appropriations and establishments for their own time, but no longer."[113] Institutions, in other words, were established not only in a universal rational time but also in a group-specific sentimental time ("their own time"). By transferring a qualitative temporal difference to the realm of politics, Jefferson opened up a new perspective on political history: in the final analysis, his argument implied the postulate of judging each temporal nation by its own standards or even of abstaining (or claiming to abstain) from moral judgments about different generations altogether.

This relativistic perspective, however, would never be fully realized in Jefferson's writings. His historical thought remained structured by the assumption of universal natural laws, even when these laws came to regulate his idea of historically diversified generations. What distinguished the 1789 letter to Madison from a more Romantic perspective on past generations–as exemplified in the succeeding year by Edmund Burke's *Reflections on the Revolution in France*–was Jefferson's inability to think of a radically contingent historical time in the singular. His artificial separation of different generations failed to embrace an uncontrollable, undivided "flow" of time from a unique past to an unknowable future; it implied a more mechanistic combination of two "separate and equal" temporalities with discernable divisions and calculable boundaries. In his construction of generational sovereignty, Jefferson thus envisioned a gradual change in time that was able to combine a universalist linearity with a degree of discontinuity, by way of locating separate generations on different, (mostly) ascending steps toward universal progress.[114] In contrast to a Burkean view of the past, which tended to stress the organic continuity of a "long succession of generations,"[115] Jefferson's gradualism established a wall of separation between the rights, duties, and temporal extension of individual generations.

His gradualist combination of two different temporalities could solve urgent political problems for Jefferson, providing the basis for his nationalization of philosophical history in the nineteenth century. His "principle that the earth belongs to the living and not to the dead" was very convenient, to begin with, in the context of the American Indians. If the earth belonged to the living majority, it no longer belonged to Logan's familial society, whose majority was dead when Logan made his speech. Nor did it belong to any other Indian nations that consisted–or, as Jefferson thought, would soon consist–merely in "remains," since their majority was already dead or would soon be dead also. Hence, it was absolutely no contradiction for Jefferson to assume that Anglo-Americans were legitimate in settling on Indian territory, while conceding to his "brothers of the Choctaw nation," for example, that "compared to you, we are but as of yesterday in this land."

For Jefferson, Indians and Anglo-Americans were contemporary generations in rational time, but they belonged to different generations in sentimental time. Hence they were, in principle, equally legitimate in their land claims, as they belonged to the same family, "born in the same land."[116]

Unless they gave up their traditional way of life, however, the Indians had what was for them a deplorable tendency to belong to the older generations of that family. And such older generations, according to Jefferson's idea of generational sovereignty, had "neither powers nor rights" over the "usufruct" of the earth.[117] As contemporaneous ancestors of Anglo-Americans, the Indians could become part of a heroic national myth, while losing all direct moral and political significance–all powers and rights–in the present. Historicizing the Indians by thinking of them as a past generation, Jefferson developed an influential rationale for the Indian policies of the next century, which would result in the cruel diminution and even extinction of many indigenous peoples. He was able, in other words, to lay a part of the intellectual groundwork for the conquest of the continent, even while emphasizing his sympathy for the Indians and his recognition of their natural equality–merely by situating them in another generational and temporal context, without having to rely, as might perhaps be expected today, on any "racist" arguments.

Jefferson's combination of two temporalities also played a crucial role in his construction of the relationship between European and American history. During his years in Paris, it enabled him to accommodate in his writings what were essentially two different interpretations of the French Revolution. As he wrote to Madison from Europe, generational sovereignty would enable Americans to exclude "the contagious and ruinous errors of this quarter of the globe."[118] For Jefferson, the privileges of Old World aristocrats still relevant all over Europe were politically as dead as the Indians populating the American continent, while the French revolutionaries were mostly "young and middle-aged" men, part of a generation that belonged more to the present and future than to the past.[119] Yet much as he liked to enlist this argument in support of the French Revolution, playing the rights of a universal present off against those of the universal past, he could also employ it to differentiate substantially between the particularities of the French and American Revolutions. Depending on the occasion and his audience,[120] Jefferson could also understand the French revolutionaries as members of a different, less progressive generation. Initiated by a different spatial *and* temporal nation, they had their particular historical conditions, including "their own time," for changing their institutions. This peculiar pace of historical change was critical for two different nuances in Jefferson's understanding of *revolution.* In the American context, he used the term quite consistently to refer to a predictable movement, subject to universal laws of nature. In the French context, however, his use of the term was ambivalent: it could signify either a well-ordered development or a chaotic occurrence that defied modes of rational control.[121]

On the one hand, Jefferson welcomed the French Revolution as "the first chapter of the history of European Liberty." Depicting it as the first of essentially similar steps in a predictable "course" of European revolutions,[122] he emphasized that the changes in France had been inspired by the American example. As he proudly observed in the summer of 1789, the American Revolution had been the "bible" for the French revolutionaries.[123] Jefferson sometimes thought the two revolutions so closely interlinked that he considered their success to be mutually influential.[124] Defending the radical turn of the French Revolution until relatively late in the 1790s, he claimed that the horrific blood toll of the Terror years was "necessary," since the "liberty of the whole earth was depending on the issue of the contest." He historicized the victims of the Terror (several of whom had been his personal acquaintances in Paris) as "martyrs" who would be posthumously recompensed by the achievement of a heroic reputation. Like that of Chief Logan, their sacrifice would enable the next generations to enjoy liberty in the future.[125]

On the other hand, however, there was also a more particularist note in Jefferson's statements about the French Revolution, even from its very beginning. As if he wanted to prepare a cautious counterargument in case the French "experiment" failed, unable to fulfill his optimistic "prognostications,"[126] he stressed from early on that particular conditions in France made the outcome unpredictable. In a letter to George Washington in 1788 he anticipated his later verdict on Marie Antoinette's regressive historical consciousness, claiming that there was a dangerous factor in French politics that the reformers had not taken into account: the destructive "influence of women in the government."[127] In the spring of 1789 Jefferson pointed out to John Jay that "habits of tyranny" were "deeply rooted" among the provincial nobility, there was danger of a "civil war," and the revolution was beginning "to wear a fearful appearance" (an expression he would reiterate in his *Autobiography*).[128] As he remarked in a more optimistic but still skeptical letter to James Madison three months later, the danger of a "temporary civil war" might be averted, but French political "practices" posed serious problems: "Never was there a country where the practice of governing too much had taken deeper root & done more mischief."[129] His metaphor of the "root" suggested an organic concept of a national character that had grown within particular spatial and temporal limits.[130] In contrast to a Romantic concept of nationhood, Jefferson thought of this national character as the result of universally comprehensible factors such as the system of government, confession, the proportion of landholders, and so on. Yet in their particular combination, these factors already produced political habits that were substantially different from American conditions. From the beginning of the French Revolution, accordingly, Jefferson emphasized in a striking number of contexts that its outcome, in his words, "cannot be foreseen."[131]

The predictive claims of philosophical history thus most obviously reached their limits in the context of Europe and its colonies. The French Revolution and its (in this form) unexpected consequences–the Terror, Napoleon's rising to power, and the independence of Haiti–suggested that "human events" were taking "their own time" rather than following the "course" of a providential nature. These problems, however, did not concern French history alone. As long as the line between the French and American Revolutions remained unclear, the threat of two divergent temporalities was felt even in America. In the late eighteenth century, the result was an eerie sense of an acceleration of time on both sides of the Atlantic.[132] When Jefferson felt that the regular "motion" of his blood was unable to "keep time with the tumult of the world,"[133] as has been said, natural history and civil history started metaphorically to diverge. Like Jenny's instantaneously graying hair in *Tristram Shandy* or Thomas Paine's exclamation "Truly, may we say, that never did man grow old in so short a time!" in 1777,[134] Jefferson's incommensurate blood circulation condensed a collective experience of "human events" that appeared to accelerate at an uncontrollable rate. The present seemed a period distinguished by the uniquely rapid tempo of historical change. As Jefferson wrote to John Adams in 1787, "I think that in the course of three months the royal authority has lost, and the rights of the nation gained, as much ground, by a revolution of public opinion only, as England gained in all her civil wars under the Stuarts."[135] And as he remarked to Maria Cosway in 1789, they had seen "in the course of fourteen years two such revolutions that were never before seen."[136] This impression of an unparalleled acceleration of historical time stemmed from the discrepancy between a rational and a sentimental time perception: whereas relatively little time had passed in a temporality measured by the earth's movement in space, a great amount of time appeared to have passed in a temporality measured by the intensity of political change. As Jefferson summarized this phenomenon in retrospect, "These years too have been like ages in the events they have engendered."[137]

In the late eighteenth century an accelerated sequence of "human events" appeared to transform the world almost too quickly to perceive and describe, let alone use them as a basis for the exemplary lessons or predictions of philosophical history. Jefferson had already observed this phenomenon in the *Summary View of the Rights of British America:* "Scarcely have our minds been able to emerge from the astonishment into which one stroke of parliamentary thunder has involved us, before another more heavy, and more alarming, is fallen on us." In this early text, however, he had still integrated "that rapid and bold succession of injuries which is likely to distinguish the present from all other periods of American story" into a restorative argument of philosophical history.[138] In his draft of the Declaration of Independence, he had similarly distinguished the present from the past by its accelerated rhythm but had continued to show himself convinced that "Future ages" would return, after the revolution, to a steadier pace of change.[139] In the last two decades of the century, however, this claim became more difficult to uphold. Much as Jefferson tried to predict an orderly future, it became apparent that events did not slow down but even seemed to speed up. As the 1790s progressed, expecting the unexpected, including in America, therefore often seemed the wiser option.

Jefferson's most frequent metaphor for this sense of chaos in the Atlantic world was that of a "storm" or "hurricane."[140] Before its accelerated movement ebbed away, it was hopeless to locate one's position in it with certainty. Jefferson was inclined to believe that the French Revolution was located at a lower stage in the progress of liberty than the American Revolution. Yet, where exactly it was located was difficult to say: at best, all nations were following the American example at a commensurate temporal distance, but at worst, they would be unable to follow it at all in the foreseeable future, possibly drawing American history back into the vortex of regressive European examples. As the century reached its close, the latter possibility did not seem unlikely. Jefferson was not alone, of course, in connecting the revolutionary turmoil to images of a catastrophic natural force–as indicated by the many metaphors ranging from volcanoes, streams of lava, and torrents of water to Robespierre's "tempête révolutionnaire."[141] According to this naturalistic imagery, civil history was not entirely separated from natural history, but was linked to extremes of rapid change in nature that, for a while, radically subverted its order. Imagining civil history as a sublime and dynamic nature, as an impersonally yet actively shaping force, clearly left behind the "boudoir" of philosophical history, which had promised a certain amount of control over (human) nature. Like natural catastrophes, events that appeared to accelerate at an unpredictable rate made this form of personal control impossible.

Jefferson's late eighteenth-century inclination to expect sudden change corresponded with his temporal horizon in the slavery question. Although he never changed his opinion that slavery was an evil and an injustice that had to be abolished, he did modify his opinion about the timing of abolition. Roughly speaking, the last quarter of the eighteenth century was the period when he came closest to expecting imminent change, when he tended to envision "human events" effecting changes in the slavery problem in the short-term future, possibly within his own lifetime. He showed himself certain that a change would happen soon, though he could not be sure what the result would be or whether change would come as part of a peaceful reform or a violent racial war.[142] In the nineteenth century, by contrast, the abolition of slavery soon became a long-term project. For Jefferson, it lost urgency as a topic, and the envisaged date of abolition measures receded into the remote future. In 1805 he made a vague long-term argument for gradual abolition, claiming that he had "long since given up the expectation of any early provision for the extinguishment of slavery among us."[143] Just a few years earlier, however, he had urged that

> the sooner we put some plan [of abolition and colonization] underway, the greater hope there is that it may be permitted to proceed peaceably to it's ultimate effect. But if something is not done, & soon done, we shall be the murderers of our own children. The "murmura venturos nautis prodentia ventos" already reached us; the revolutionary storm, now sweeping the world, will be upon us, and happy if we make timely provision. From the present state of things in Europe & America, the day which begins our combustion must be near at hand; and only a single spark is wanting to make that day to-morrow. If we had begun sooner, we might probably have been allowed a lengthier operation to clear ourselves, but every day's delay lessens the time we may take for emancipation.[144]

Written under the fresh impression of the revolution in Saint Domingue, where the abolition of slavery during the second year of the French Republic had been followed by violent racial conflicts between blacks, mulattoes, and whites (and the forced expulsion of the latter), this was one of Jefferson's most emphatic calls for the rapid initiation of a gradual abolition plan. To be sure, apocalyptic scenes were part of a conventional repertory in the enlightened antislavery discourse,[145] which Jefferson may have employed to adapt his views to those of the recipient of the letter.[146] Nevertheless, the imminence of apocalyptic change in this letter ("only a single spark is wanting to make that day to-morrow") reveals a heightened sense of alarm. It is remarkable that Jefferson seems to have expected the "revolutionary storm" inevitably to hit America in the near future: the storm itself could not be avoided; only "timely provision" could be taken. In this letter it seems that in order to end the vicious circle of exemplary history and prevent Virginia slaves and free blacks from emulating the violent example of their counterparts in the French colony, American citizens had to act. If they failed to prepare themselves for the storm, they would become "murderers" of their "own children," thus directly inverting the principle of generational sovereignty.

Unfortunately, however, statements like these did not eventually induce Jefferson to advocate a speedy beginning of gradual abolition plans in the United States or Virginia, even after Gabriel's Conspiracy in 1800 had suggested that his apprehensions of impending slave rebellions on the mainland had not been unjustified. Instead of political action on the slavery question, he merely increased the efforts of his "masterly pen," trying to make the most of the situation rhetorically and using the epistemological instability of the *fin-de-siècle* crisis in his favor. Thus taking advantage of the partial breakdown of universalist modes of historical explanation, he sought to construct an exceptionalist narrative of American history, a history that was not only "fundamentally different" from that of European antiquity but also "materially different" from that of European modernity. To this end, he sought to distribute the two temporalities contained by generational sovereignty in space, trying to persuade himself and others of a clear-cut break with the European past. Americans and Europeans belonged not only to different spatial nations but also to different temporal nations (or generations), which had to act separately "in their own time." The crisis of the universalist lessons, judgments, and predictions of philosophical history thus also had a liberating effect for Jefferson: one way of dealing with the threat of a repetition of French and Caribbean examples in the United States was altogether to deny their exemplary function.[147] In Jefferson's construction of a difference of historical quality between Europe and its colonies, on the one hand, and the United States, on the other, philosophical tensions within his historical consciousness thus merged with his actual historical experience and political needs.

The party conflicts of the 1790s gave Jefferson more and more political reasons to differentiate substantially between European and American history. These controversies, which had begun over Hamilton's economic program, gained greater emotional intensity due to the dissent over the radical turn of the French Revolution and the heightening diplomatic tensions with the French republic, which very nearly resulted in an open war.[148] In what was largely still a premodern understanding of a political partisanship, the very existence of two parties appeared to many contemporaries a sign of decline, foreboding a disastrous future for the republic. Occasionally therefore, the heated discussions in the House of Representatives and the Senate were themselves likened to the noise of a "Tempest" or "Hurricane."[149] For someone reputedly as averse to personal conflict as Jefferson, it must have been an especially stressful experience to act as one of the leading figures in these political controversies,[150] all the more so as the tempest of political discord itself did not look good on the historical record. In this atmosphere of impending doom, the emerging party leader Jefferson employed two strategies to reconcile his partisanship with his hopes for a stable republican future. One strategy was, of course, to deny the existence of parties and to describe recent political conflicts as a problem of epistemology rather than of politics. (As he often argued, the majority of Americans were "substantially republican," and the rest had a problem, basically, with their eyesight.)[151] And the other strategy he began to develop was to argue that in general the negative examples of political disasters in European history were not directly applicable to American history.

For this argument, Jefferson constructed a homogeneous American history, which he then sought to distinguish fundamentally from European conditions.[152] Instead of claiming that the French revolutionaries had followed the American example, it became politically important in the later 1790s to emphasize another exemplary historical connection, located solely *within* the American context. Jeffersonians began to construct a link between the crisis of the 1790s and the crisis of the 1770s, depicting themselves as the true heirs of the American revolutionaries two decades earlier. The figure of Jefferson became the most conspicuous link in this rhetorical chain between the American "spirit of 1776" and the American "principles of ninety-eight."[153] Especially when he was confronted with Federalist accusations of being a secret missionary of the French Revolution, it became personally important for Jefferson to distinguish decisively between American and French politics–thus moving further away from his former universalist approach. Commenting on a second "great revolution" in France in 1800 (the eighteenth Brumaire), Jefferson claimed that the French and the American Revolutions were "materially different." Even when writing to a fellow Republican, he voiced his "fear of the influence of the [French] example on our countrymen. It is very material for the latter to be made sensible that their own character & situation are materially different from the French, & that whatever may be the fate of republicanism there, we are able to preserve it inviolate here: . . . Our vessel is moored at such a distance, that should theirs blow up, ours is still safe, if we will but think so."[154]

A year before a turning point in American history that he would praise, in retrospect, as the "Revolution of 1800," and half a year before an attempted "revolution of 1800" among Virginia slaves that he would usually prefer not to mention later, Jefferson developed an argument that would characterize many of his nineteenth-century writings. Repeating the word in this letter, he drew a "material" line between European and American history. His nautical metaphors not only took up the revolutionary storm metaphor of the 1790s, stylistically situating the contrast between the continents in the domain of recent maritime conflict. They also evoked the "material" geopolitical fact of the Atlantic Ocean, advantageously separating the young American republic from the chaos of European history. The imploring tone of the last words–"if we will but think so"–paradoxically suggested, however, that the distance between the "materially different" French and American historical conditions had not yet "materialized" in Jefferson's thought. It was still mainly an article of faith that he could only hope would influence American politics favorably. In the following year, this imagined distance finally became "material." Jefferson's first inaugural address continued his nautical imagery, calling for support "to steer with safety the vessel in which we are all embarked amidst the conflicting elements of a troubled world." Yet in contrast to in his last year as vice president, Jefferson now spoke of the Atlantic Ocean not as an intellectual construct but as a purposeful natural fact that "an overruling Providence" had benevolently arranged for Americans: "Kindly separated by nature and a wide ocean from the exterminating havoc of one quarter of the globe, . . . what more is necessary to make us a happy and a prosperous people?"[155]

In Jefferson's nineteenth-century reflections on history, this fundamental divide between European and American history, both material and ideal, became a ruling theme. He tended to differentiate substantially between the "exterminating havoc" of Europe on one side of the ocean and the peaceful prosperity of America on the other. In this manner, he attempted to arrange his two revolutionary temporalities in space. Leaving the chaos of sentimental time in Europe, he tried, more or less successfully, to associate the order of rational time with the United States. In an argument that corresponded to his criticism of European aristocrats on the individual level of time perception, Europe became the domain of "history," in the negative sense of political "crimes & calamities." America, by contrast, became the domain of a hopeful republican future. Thus, Jefferson's spatial organization of generational sovereignty became his contribution to a larger cultural tendency to transfer seventeenth-century forms of exceptionalism to the context of a secular American progress.

In the autobiographical narratives Jefferson wrote in his retirement, he accordingly conveyed very different images of the French and American Revolutions. He moved away from defending the radical turn of the French Revolution as necessary for universal progress and from considering the Jacobins, as he had in 1792, "as representing the true revolution-spirit of the whole nation."[156] Instead, he depicted European history as a period of accelerated moral decline rather than progress, a chaos characterized by "rapid depravity" and "disgrace," "horrors," "atrocities," "murderous Jacobins," and "the total extinction of national morality."[157] To this he opposed an American history that had slowed again to a reliable pace of progress. Due to the "steady and rational character of the American people,"[158] the United States was the first country to look forward to the gradual realization of a "republican millennium."[159] In Jefferson's old age, when he saw the future of his younger years–the American Revolution and the founding of the American republic–recede into an increasingly distant past, he could still think of this past as a hopeful future: Europeans, South Americans, and eventually all nations would struggle to become republics, as Jefferson saw it, by trying to emulate the "example" of American history. In his eyes, the history of the early American republic was thus becoming the republican future for the rest of the world.

This modified recourse to exemplary modes of historical explanation marked an important development in Jefferson's thought. With his election to the presidency, he had Americanized the philosophical history of the Enlightenment, making its positive didactic potential an exclusive attribute of the national American context. In his views of American progress, the rhetorical function of history began again to dominate, accounts of the past becoming literary devices to achieve future political goals. For his claim that the national history of the young United States was the only progressive example in a "troubled world,"[160] of course, Jefferson had to be quite selective as to which parts of his historical experience to represent. Thus, his consolidated teleology of American history had a significant victim: his demands for "timely provision" in the slavery question, as still expressed in 1797. After his election to the presidency, Jefferson made great rhetorical efforts to claim that a "hurricane" might have come dangerously near the American coast but had eventually failed to make landfall. As he tried to see the situation in the early nineteenth century, unpredictable "human events" continued to characterize the history of a war-torn Europe and its colonies but could have no significant effect on the progressive order of American history (an argument that became more convincing after the Louisiana Purchase). In the wishful thinking of Jefferson's presidential optimism, it was crucial that the continued existence of American slavery appear as a temporary part of a providential order rather than as a subversive element drawing Americans back into the European turmoil.

The most elegant mode of achieving this rhetorical goal was to refrain from mentioning slavery in general, especially from mentioning it as a problem that required a solution in the short-term future. Within his Americanized philosophical history, Jefferson could try to show himself as certain that the abolition of slavery would be a controlled long-term measure rather than a chaotic and potentially violent event. As he emphasized the "steady and rational character of the American people," he could also convince himself that all Americans would act rationally, at least in the long run. They would display the future-orientation inherent in their national character by their steady pursuit of gradual abolition in all the states. In the ordered universe of philosophical history, the nineteenth-century Jefferson could try to see historical evidence of a hopeful future in two developments. Firstly, the abolition of the Atlantic slave trade seemed to promise an end of slavery: eliminating the "cause" of the trade would gradually also eliminate the "effect" of slavery, as if another evil Old World influence on American life simply had to be cut off. And secondly, Jefferson's renewed faith in universal order enabled him to see parallels between spatial and temporal developments in the disappearance of slavery: since abolition had begun in the northern states, where the slave population had been proportionally smaller, it could be expected to progress, in different stages, toward the south, until it would finally reach the most southern states, which had the greatest proportion of slaves and the slowest rate of change.

Accordingly, the nineteenth-century Jefferson changed his position on slavery in the western territories. In contrast to his views in the 1780s, he no longer regarded the expansion of slavery as a problem. Since only a relatively small number of slaves would be "diffused" over an immense territory, slavery could be expected to disappear, in the spatial and temporal "order of events," even more quickly than in the northern states, whose extension was diminutive compared with that of the West. The exact point in time when all American slaves would eventually be free—and when, Jefferson hoped, they would have left the country, since they had the potential of being a different "nation"[161]–could not be known from a limited human perspective of the providential plan of the gradual disappearance of slavery, but the benevolent "design" of the plan itself seemed certain. In Jefferson's Americanized philosophical history, the abolition of slavery could recede into a distant future also because, within a rational time perception that abstained from qualification, the exact point in time was insignificant compared with its orientation toward a larger future goal. As Jefferson saw it, therefore, the priority for American citizens could not be a speedy abolition of slavery. Rather, it had to be the effort to keep the American union of republican states intact, since this union alone could provide a positive historical example for the rest of the world.

Chapter 6

Beyond Example?

On New Year's Day, 1812, John Adams wrote to Jefferson and ended a silence that had subsisted between them since the beginning of the century. Their ensuing retirement correspondence offered the two former presidents the possibility not only to discuss a wide range of political, historical, and philosophical problems but also to convey some of the most congenial impressions of their characters. Occasional disclaimers notwithstanding, both elder statesmen also had posterity on their minds when celebrating the renewal of their friendship above party politics. Their correspondence may thus be viewed as their most successful autobiographical construction, one that turned out to be more complimentary to their posthumous reputations than their actual autobiographies. As if to gain clarity about their historical legacies at the onset of this unique exchange, Jefferson in his first reply to Adams summed up their place in his vision of American history. After a short essay on household manufactures occasioned by a typical misunderstanding of Adams's wordplays, Jefferson reflected on "recollections very dear to my mind" of the revolutionary past, when the two men had been "fellow laborers in the same cause."[1]

As in the very last letter of their correspondence, he dressed his reminiscences of their "Heroic age"[2] in the nautical metaphors familiar from his first inaugural address: "Laboring always at the same oar, with some wave ever ahead threatening to overwhelm us and yet passing harmless under our bark, we knew not how, we rode through the storm with heart and hand, and made a happy port." Although the young United States was at the brink of a new war when he wrote this sentence in 1812, Jefferson chose to emphasize that the American ship of state had managed to escape the dangerous European tempests and Caribbean hurricanes of the recent past: aided by Providence, he and Adams had collaborated successfully to ensure the ship's safe progress through the "conflicting elements of a troubled world."[3] In an Atlantic crossing from the east to the west, as it were, they had left the turmoil of European history behind and "made a happy port" in America.

Proceeding from the past to the future, Jefferson now showed himself entirely convinced of American progress:

> So we have gone on, and so we shall go on, puzzled and prospering beyond example in the history of man. And I do believe we shall continue to growl [grow], to multiply and prosper until we exhibit an association, powerful, wise and happy, beyond what has yet been seen by men. As for France and England, with all their pre-eminence in science, the one is a den of robbers, and the other of pirates. And if science produces no better fruits than tyranny, murder, rapine and destitution of national morality, I would rather wish our country to be ignorant, honest and estimable as our neighboring savages are.[4]

Jefferson thus reemphasized his conviction of a fundamental divide between American and European history. In the United States, the future was predictable through the knowledge of a similar past ("So we have gone on, and so we shall go on"), allowing Americans to continue in their steady progress toward the future of a republican millennium ("an association, powerful, wise and happy, beyond what has yet been seen by men"). The first of all nations to have this peaceful republic within reach, Americans were "puzzled and prospering beyond example in the history of man"–"puzzled" by a providential plan that was only partly accessible to a limited human knowledge of history. Although they "knew not how," Americans were moving "beyond example in the history of man" because, Jefferson hoped, their progress avoided the mistakes of a contingent European history. Americans were themselves becoming the only positive historical "example," the republican model to be emulated by all other nations in the time to come. Thus, while American history was indeed moving "beyond" the examples of European history, Jefferson's theory of history did not move "beyond example," as he reaffirmed the didactic function of history: in the case of the American past, he tried his best again to understand history as "philosophy teaching by examples."

Claiming to prefer the original innocence of the Indians to the moral destitution of French "robbers" and English "pirates," Jefferson even alluded to the restorative dimension of his early philosophical history. To be sure, this primitivism cannot be taken at face value, as he was usually convinced that science could indeed produce "better fruits" than tyranny and murder. However, he could make good use of this point in his Americanization of philosophical history. The rhetorical conflation of Anglo-Americans and their western "neighbors" into the same moral nation fitted well into his dualistic construction of universal history: in the heterogeneous temporality on the eastern side of the Atlantic, the virtuous beginnings of the French Revolution had resulted in the moral opposite of Napoleon's "robberies," whereas, in the homogeneous temporality on the western side, the natural morality of the Indians was continued among Anglo-Americans (at least if they took care, as Jefferson did, not to mention the "piracies" and "robberies" committed by American slave traders and slaveholders).

Jefferson also returned to a rational temporality and to philosophical history on another level in this letter: he claimed to have "given up newspapers in exchange for Tacitus and Thucydides, for Newton and Euclid," and to find himself "much the happier."[5] Establishing what would become a frequent motif of his retirement correspondence, Jefferson stylized his personal life as a withdrawal from the politics of the day. Instead of the short-term import of "lying" newspapers,[6] he turned to the timelessly valid writings of the ancient founding fathers of philosophical history and foundational thinkers of natural philosophy and mathematics.[7] Thus, he transferred the divide between two temporalities also to his own situation as elder statesman. From the detached perspective of a rural retirement, Jefferson tried to see the remainder of his life as the calm succeeding the storm. When the rhythm of his years slowed down, he liked to think of himself as an American sage taking the long-term view, serenely remembering the beginnings of the American republic, while looking forward to his posthumous reputation as one of its principal founders. This "classical" Jefferson was visualized in the portrait Jefferson commissioned from Gilbert Stuart "in the medallion form"–a sketch *à l'antique* Jefferson considered such a "fine thing" that he paid the artist "another 100 dollars, probably the treble of what he would have asked" (figure 6).[8]

Unfortunately, however, Jefferson's personal vision of his retirement turned out to be as difficult to realize as his political vision of a linear American progress. In his reply, Adams mentioned a central problem in his friend's philosophical history: "Your Memoranda of the past, your Sense

Figure 6. Thomas Jefferson, by Gilbert Stuart, 1805 (the medallion portrait). (Harvard Art Museums, Fogg Art Museum. Gift of Mrs. T. Jefferson Newbold and family, in memory of Thomas Jefferson Newbold, class of 1910, 1960.156. Photograph: Imaging Department, © President and Fellows of Harvard College)

of the present and Prospect of the Future seem to be well founded, as far as I can see. But the Latter i. e. the Prospect of the Future will depend on the Union: and how is that Union to be preserved?"[9] As Jefferson had to realize as well, human contingencies threatening to impair the American union continued to cause many problems, complicating his narrative of progress more than he often liked to admit. In moments of political turmoil and personal depression–most of all, during the Missouri Crisis–he had to struggle, again, with his skepticism about the predictability of history, even in the American context.

In this chapter I address the ambivalence that defined both the content and the method of Jefferson's Americanized philosophical history in his later years. I begin with his presidential and postpresidential emergence from the historical skepticism of the 1780s and 1790s. In contrast to a number of European thinkers, whose doubts about the didactic function of history were becoming more radical, Jefferson tended to speak of historical "lessons" and "examples" with a renewed intensity in the nineteenth century. Due to an increasing public interest in the biographies of the revolutionaries in his retirement, these new American examples even tended to be classical examples of "great men" again, not least, of course, of Jefferson himself. As he tended to display a greater conviction of the positive moral potential of American history, he also became more pragmatic in regard to the methods of political historiography. Lest the American Revolution be misrepresented to posterity by Federalist writers (conveying falsely regressive lessons to later generations), Jefferson subdued his skepticism about historiography as a genre and sought to construct, as a necessary supplement to his collection of primary documents, his own philosophical history of American progress.

This narrative, however, was no simple return to the moralizing "philosophy teaching by examples" of his youth. Jefferson's concept of history had become more complex, and the Aristotelian hierarchy of poetry and historiography that had dominated his early writings was almost reversed: rather than arguing, as he had done half a century earlier, that the fictional examples in literature were superior to those of real life recorded by history, the older Jefferson explicitly denied the didactic value of most fictional writings, criticizing them for their pernicious neglect of "reason and fact."[10] In contexts when the "post-revolutionary" generation of Americans did not fulfill Jefferson's expectations, moreover, the spatial division between European and American history became permeable. In these moments, he described the American situation in European terms and left the country's future an open question. As Jefferson's nationalized philosophical history combined elements from diverse historical sensibilities, the ahistorical concept of "philosophy teaching by examples" survived in a worldview that was becoming increasingly historical.

A Clock Wound Up Again

The event that Jefferson proudly called the "Revolution of 1800" became a cornerstone of his Americanized philosophical history.[11] By referring to his precarious but eventually peaceful election to the presidency in 1801 as "Revolution," he defined the term as an orderly movement instead of an uncontrollable eruption. What remained from the more modern concept of revolution that had sometimes resonated in his late eighteenth-century writings was a sense of the uniqueness and "newness" of the situation, after

America had supposedly acquired, by a peaceful transition of power, "the strongest Government on earth."[12] Jefferson's expression "Revolution of 1800" conflated the event of his narrow election in 1801 with the universally more impressive turn of the century in 1800. It blended the contingency of unpredictable "human events" with the order of a universal "course" of history, fusing a reference to a temporality measured by the significance of "human"–social and political–change with a reference to a temporality measured by the earth's "natural" movement in space and the number of full centuries elapsed since the birth of Christ. Thus, it endeavored to supply the event of Jefferson's election to the presidency, accidental as it must still have seemed in early 1801, with a quasi-providential necessity. In a famous letter to Joseph Priestley in March 1801 Jefferson described the ending of the "revolutionary storm" and the beginning of his first term as president with powerful religious overtones:

> As the storm is now subsiding, and the horizon becoming serene again, it is pleasant to consider the phenomenon with attention. We can no longer say there is nothing new under the sun. For this whole chapter in the history of man is new. The great extent of our Republic is new. Its sparse habitation is new. The mighty wave of public opinion which has rolled over it is new. But the most pleasing novelty is, it's so quickly subsiding over such an extent of surface to it's true level again. The order & good sense displayed in this recovery from delusion, and in the momentous crisis which lately arose, really bespeak a strength of character in our nation which augurs well for the duration of our Republic; & I am much better satisfied of it's stability than I was before it was tried.[13]

Dismissing the cyclical temporality of Ecclesiastes ("there is no new thing under the sunne"), Jefferson emphasized that a significant part of the secular newness he perceived after the election of 1800/1801 was the order of political change in the United States. He illustrated this idea by the metaphor of a "mighty wave" rolling over a "great extent" of territory and quickly receding to its natural level afterwards. Evoking a secularized version of the biblical deluge, this image implied a change that conformed to natural laws, in a temporality that could be measured by an external movement in space. This view of American progress left behind influential European traditions, such as Hellenistic assumptions about time in the cyclical temporality of Ecclesiastes and classical republican notions of space in Montesquieu's small-territory theory.[14] The peaceful circumstances of political change appeared to Jefferson the best empirical proof of natural order in American history. Even in the worst case of a "non-election" in 1801, when the "government would have been in the situation of a clock or watch run down," as he told Priestley, the "clock" of the American government would not have stopped forever, since Americans would always have "wound it up again."[15] Like the metaphor of the "wave," the "clock" described a temporal sequence in American history as something that was spatial as well as visual. It emphasized that Jefferson again expected Americans, after a chaotic interval of "delusion," to perceive change in time by visual impressions of a movement in space rather than by abstract processes in the mind. With the "Revolution of 1800," American citizens had supposedly regained their clear empirical vision of reality,[16] and with it their linear future-orientation.[17] The "order & good sense" in their national character appeared to be the best safeguard for the future "duration" of their republic. If Americans now "wound up the clock," they performed this symbolic act of re-creation, not in a Shandean sense, but with the serious purpose of establishing a more perfect order in their clockwork universe.

Adams felt insulted when he read a published version of this letter in 1813, since Jefferson had explicitly cited him ("the president himself") among those who wished "to look backwards, not forwards" in the education of American citizens. In one of several letters venting his indignation, Adams also took issue with Jefferson's spatial metaphors of the "wave" and the "clock," exclaiming:

> Oh! Mr. Jefferson! What a Wave of public Opinion has rolled over the Universe? By the Universe here, I mean our Globe. I can yet say there is nothing new Under the Sun, in my Sense. The Reformation rolled a Wave of public Opinion over the Globe, as wonderful as this; A War of thirty Years, was necessary to compose this Wave. The Wars of Charlemaigne rolled a Wave. The Crusades rolled a Wave, more mountainous than the French Revolution. Only one hundred Years ago, a Wave was rolled; when Austria England and Holland in Alliance, contended against France, for the dominion or rather the Alliance of Spain. Had "The Clock run down," I am not so sanguine, as you, that the Consequence would have been as you presume.[18]

For Adams, who had lost the elections of 1800 to Jefferson and Aaron Burr, the "mighty wave of public opinion" washing Jefferson into the presidency had been neither exceptional and "new" nor "mighty" and providentially ordained. In the comparison to the numerous other "waves" of world history, Jefferson's election appeared insignificant, especially as its narrow result had depended, as Adams reminded his former competitor, on a contingent situation in the House of Representatives that could technically have resulted in Burr's victory as well. As Adams's criticism illustrates, the providential exceptionalism of Jefferson's version of American progress was anything but an uncontested concept in his generation. How Jefferson himself would have managed to uphold his progressive expectations on American history if the election of 1800/1801 had brought about a different result remains an interesting question.

Such argumentative weaknesses, however, only increased the rhetorical fervor in Jefferson's efforts to draw a providential line between European chaos and American order. Writing to his friend Alexander von Humboldt in the year of Adams's letter, for instance, Jefferson justified the beginning War of 1812 by America's nature-ordained difference from the "never-ceasing broils of Europe." For the difficult justification of an American war fought in order to be different from a war-torn Europe, he had to evoke his typical blend of a descriptive and a prescriptive understanding of "nature" in a homogeneous temporality, referring to the "insulated state in which nature has placed the American continent" as a political argument for the expulsion of European interests from the continent. Their providential geopolitical situation allowed Americans–or at any rate should allow them in the future–a situation in which "no spark of war kindled in the other quarters of the globe should be wafted across the wide oceans which separate us from them."[19] When Jefferson wrote these lines in 1813, the "spark of war" had actually also been wafted from the United States across the Canadian border. Yet in contrast to European nations, as Jefferson saw it, American republicans were not enforcing an expansionist policy, but were only trying to realize the political equivalent of their "natural" American hemisphere. The War of 1812 was a war fought against European wars, meant to achieve the full realization of the providential difference between American progress and the history of European "wars & contentions."[20]

In this logically fragile train of reasoning, the fate not only of the American republic but of the whole world was at stake. For Jefferson, universal progress depended on the success of the American experiment and the rhetorical demonstration of its difference from the martial culture of Europe. Unfortunately, in order to secure this positive example and illustrate the peaceful potential of the United States, actual wars had to be fought. As he told his former secretary William Short, the English enemy had the "consolation of Satan on removing our first parents from Paradise: from a peaceable and agricultural nation, he makes us a military and manufacturing one."[21] Jefferson's solution to this problem was provided, as in the case of the "first parents," by a quantifying concept of change in time. When Americans had to face a political crisis in the present, the republican millennium that he had sometimes thought near at hand (as in the above letters to Adams and Priestley) was merely postponed to a more remote time to come.[22] For non-Americans following the American example, it was accordingly located in a yet more remote future. "African" slaves would have to wait still longer for their emancipation and the founding of their own republic. Europeans likewise had to display the virtue of patience: the temporal horizon of Europe's redemption from despotism could recede into an indeterminate future in Jefferson's old age. By 1816 he saw "an awful void between the present and what is to be the last chapter of that history," a void that would be "deluged . . . with rivers of blood which are not yet run out." As a compensation for this European version of the deluge, Jefferson claimed that the moral "object" of secular history, the "general amelioration of the condition of man" in a world of peaceful republics, was determined and invariable. For him, it was "fixed in the eye of nations."[23]

In one of his last letters, Jefferson reemphasized that the history of the American Revolution would eventually ensure that "all eyes are opened, or opening, to the rights of man."[24] American history thus became a new Ciceronean *magistra vitae,* part of a rhetorical argument for universal progress. This argument relied not only on a considerable amount of wishful thinking but also on a quantifying rather than qualifying temporality. In the letter to Humboldt, for example, Jefferson tried to gloss over the problem of the "awful void" between the present and the future by dedicating a whole paragraph to his view that "fifty years are soon gone over." If the aim of world history was known, fifty years more or less of war or slavery did not matter. What mattered was that all nations kept their eyes "fixed" on the moral aim of history. In order to ensure this "fixation," American history had to become a new "philosophy teaching by examples," preventing all nations from losing sight of the hopeful lessons taught by the American republic.

As Jefferson was aware, however, the "magic" of a masterful style was necessary to supply the events even of American history with a future-oriented structure. This problem moved to the foreground during his presidency, when he began to reflect on new means to protect his view of the American Revolution against competing Federalist interpretations. His skeptical escape from the master narratives of philosophical history—his alternative focus on the antiquarian activities of collecting and preserving primary documents–hardly appeared sufficient to promote Republican interests in what became the "biographical war" of the early nineteenth century. Although John Marshall had used Washington's papers in his popular *Life of George Washington* (1804–7), Jefferson nevertheless detested the work as a Federalist "party diatribe."[25] The primary material that Washington's heir had opened to the chief justice, far from correcting his Federalist bias, had only lent additional credibility to his interpretation. It almost appeared posthumously to authorize the partisan view of Washington's presidency in the fifth volume of his *Life of Washington,* an interpretation in which, predictably, Jefferson and Jeffersonian Republicans fared rather badly.

This experience may have reinforced Jefferson's pragmatic emphasis on the need for a didactic historiography that relied on the literary "design" of the narrative. Due to an undiminished concern about his republican integrity, however, he at first tried to avoid the challenge of writing this exemplary history himself. Unlike John Adams, he hesitated personally to engage in the "biographical war."[26] Instead, he searched for an appropriate writer able to relate the "true" history of the American republic. During his presidency, he at first tried to enlist his friend Joel Barlow, offering him the use of Madison's papers and his own. But Barlow preferred to deal with the American Revolution in the genre of epic poetry, a genre that Jefferson, in contrast to his earlier views, no longer regarded as an equally useful instrument of republican education.[27] After this and other attempts to encourage and supervise a Republican history of America had failed, and Federalist writers always seemed to be ahead in the historiographical arms race,[28] Jefferson began more actively to prepare his own contribution to American historiography: a careful selection of papers combined with two larger historiographical narratives.

To be sure, when he was approached by outsiders asking him to write a Thucydidean history of the political events in which he had been engaged, Jefferson would still disclaim any interest in doing so.[29] His historiographical activities largely took place behind the scenes, in a realm of half-secrecy that only a few friends and allies were allowed to enter.[30] Perhaps inspired by Washington's example, he wished his writings to be published only after his death, when they could best convey his personal disinterestedness. The retired Jefferson constructed the persona of the serenely detached elder statesman whose perfect conscience allowed him passively to await a secularized Last Judgment, the estimate of future historians at the "bar of posterity."[31] This pose was most conspicuous in his exchange with the more overtly emotional Adams, from whom Jefferson tried to distinguish himself in statements like the following: "As to myself, I shall take no part in any discussions. I leave others to judge of what I have done, and to give me exactly that place which they shall think I have occupied. Marshall has written libels on one side; others, I suppose, will be written on the other side; and the world will sift both, and separate the truth as well as they can. I should see with reluctance the passions of that day rekindled in this, while so many of the actors are living, and all are too near the scene not to participate in sympathies with them."[32]

While the last sentence, stating his aversion to personal conflict (most of all, with Adams himself), may have been sincere, Jefferson's implication that Republican histories would be on the same level as Marshall's Federalist "libel" was probably less so. This letter illustrates that he had at his disposal a concept of historical "impartiality," according to which historical "truth" had to be found somewhere in the middle between Federalist and Republican interpretations. Yet, as many other statements indicate, the aged Jefferson was indeed hoping for a powerful Republican "libel" to be written. His pose of placidly awaiting the judgment of posterity did not prevent him from making great efforts to induce future historians to give him "exactly that place" which *he* thought he had occupied. After he had "made history" in the sense of shaping the historical events themselves, Jefferson continued in this activity, concentrating on the making of his Americanized philosophical history.

His two larger autobiographical narratives, the 1818 explanations to the *Anas* and the 1821 *Autobiography,* can be read as a continuous narrative of his life. Instead of these titles, which were used in later editions, Jefferson referred to his political memoirs loosely as "memoranda" or "memorandums,"[33] suggesting their future use as aids to the (personal or national) memory. Together, they can be understood to form "a coherent, didactic history of the Revolution and early republic,"[34] not only because they cover the period from Jefferson's birth in 1743 until his second administration but also because of their shared political function: both works depict Jefferson as the exemplar of a virtuous republican in a crucial period of world history. The *Autobiography* presents him as the author of the Declaration of Independence, as a sympathetic but detached observer of the beginning French Revolution, and as an opponent of slavery. And the main point of the *Anas* was to convey an image of Jefferson as an honest public servant and loyal friend of George Washington's during the 1790s. Since these were contested interpretations in Jefferson's generation, he was greatly in need of a philosophical history supposed, first of all, to make *himself* a "better man, and better citizen" in the eyes of posterity.

To achieve this aim, the *Autobiography* and the explanations to the *Anas* had to confront complex literary problems, problems that, arguably, they did not always solve. In their efforts to convince readers of American progress, they mostly stressed historical necessity at the expense of contingency. Events were understood to follow a providential "design," but a design that was accessible only in part by a limited human understanding. Jefferson's narrative directly reflected this limitation. For instance, he characterized the impact of the American Revolution on world history as "a wonderful instance of great events from small causes. So inscrutable is the arrangement of causes & consequences in this world that a two-penny duty on tea, unjustly imposed in a sequestered part of it, changes the condition of all it's inhabitants."[35] Thus, Jefferson showed himself convinced of a teleological universe, the details of whose "chain of causes and effects," however, remained as deeply concealed from the historical observer as the Saturday of the Great Clock from a casual visitor to Monticello.

The modest assumption of a partial access to a larger historical causality influenced the style, narrative structure, and concept of character in the *Autobiography*. Much longer than the explanations to the *Anas,* it reads like the unassuming list of "recollections of dates & facts" announced in the first sentence. The text differs from many of Jefferson's other writings in its relatively unsophisticated syntax. Long passages are characterized by an abrupt staccato rhythm that contrasts strangely with the more elegant flow, for instance, of his draft of the Declaration of Independence, which he inserted into the text. The paratactic structures reflect the simplified notion of historical causality to which he confessed. Reminiscent of the lists and charts containing data of natural history in *Notes on the State of Virginia,* they gave his "memoranda" the form of a narrative enumeration, supporting few attempts to offer more ambitious historical explanations. Interrupted only by occasional side remarks presenting universalizing moral lessons (which are usually characterized by a more fluent diction, as in the quotation above), the *Autobiography* presents a list of events for the most part unencumbered by efforts to look beneath the surface. As the details of the universal "arrangement of causes & effects" were ultimately "inscrutable," precise information on this arrangement was superfluous. What mattered was not so much historical explanation as an assertion of the "fixed" moral aim of world history.

Within the bounds of his providentialism, Jefferson's major mode of historical explanation remained personal and voluntaristic: he tended to assume that "human events" were planned or obstructed by more or less rationally acting individuals.[36] Depending on the universal moral value of their "principles" reflected in their actions, individuals contributed either to accelerating or to postponing the republican millennium. Hence, it was a central function of Jefferson's "memoranda" to pass moral judgments on its characters. His account of the 1770s, for instance, introduces James Madison thus:

> Mr. Madison came into the House in 1776. . . . Never wandering from his subject into vain declamation, but pursuing it closely in language pure, classical, and copious, . . . he rose to the eminent station which he held in the great National convention of 1787. . . . With these consummate powers were united a pure and spotless virtue which no calumny has ever attempted to sully. Of the powers and polish of his pen, and of the wisdom of his administration in the highest office of the nation, I need say nothing. They have spoken, and will forever speak for themselves.[37]

When Jefferson introduced a new character into his *Autobiography* or into the *Anas,* he tended to attach a character sketch, a timeless moral "portrait" of the person's entire life rather than of his or her life up to the point of the event described. Thus moving clumsily (to modern sensibilities) back and forth in time, Jefferson destroyed most possibilities of curiosity or suspense. With few exceptions, he abstained from dramatizing key moments of his personal or political life by references to an unpredictable future.[38] In a homogeneous temporality whose past, present, and future were focused alike on a "fixed" historical telos, it was possible for Jefferson to proceed anachronistically in the larger construction of his narrative as well. Curiously similar, in this respect, to David Hume's "Tory" work of philosophical history, the *History of England,* Jefferson's autobiographical memoranda were composed in reverse order. Starting in 1818, with an account of the years 1791–1806 in the *Anas,* he completed the narrative of his life in 1821, with the period 1743–90 in his *Autobiography.* Also as a result of this anachronistic procedure, he gave almost no account of his own emotional and intellectual education.[39] The figure of Jefferson essentially remains as "fixed" as Madison's once-and-for-all character portrait at the moment of his first appearance in the text.

From today's perspective, this lack of character development and of a specifically human historicality makes Jefferson's history appear closer to a "chronicle" (a term that for him already implied a criticism) than an autobiography.[40] It is no coincidence, therefore, that it has been analyzed more for its political content than for its literary merits. In particular, its absence of character development has led to a widespread disregard for this work by literary historians and critics, who tend to draw unfavorable comparisons with Benjamin Franklin's earlier but more "modern" autobiography. Jefferson's memoirs, oddly "unconstructed" and carelessly written as they appear, mostly fail to produce the vivid impressions achieved by the autobiographical persona in Franklin's text or even by some of Jefferson's own personal letters. A reason for this lack of literary appeal can be found in a central paradox of his *Autobiography:* as the above sketch of Madison illustrates, the text tends to proclaim itself redundant. Ideally, at the universal "bar of posterity" the actions of a man "forever speak for themselves." It was only in less ideal situations in a provisional present, when the republican millennium was postponed due to Federalist distortions of the Republican "truth," that texts like Jefferson's history became necessary. Like the explanations to the *Anas,* the *Autobiography* therefore is best read as an auxiliary instruction on how to understand his selection of notes, letters, and documents. Often referring directly to his original papers,[41] Jefferson's historical work presents itself as a supplementary device in the larger rhetorical aim of public education.

It remains an interesting question whether Jefferson himself thought that his history would be successful as a "designed" work of art, fulfilling its didactic function by the aid of his masterful style. His apparent listlessness in composing the *Autobiography,* combined with his frequent references to the great effects achieved by stylistic masters of history (from Thucydides and Tacitus to Hume), might suggest a certain frustration in regard to his personal capabilities in the field. In his relatively more emotional (and stylistically somewhat more experimental) explanations to the *Anas,* Jefferson included the following criticism of a historical work:

> The partiality of this pen . . . shews itself too in the cold indifference with which a struggle for the most animating of human objects is narrated. No act of heroism ever kindles in the mind of this writer a single aspiration in favor of the holy cause which inspired the bosom, & nerved the arm of the patriot warrior. No gloom of events, no lowering of prospects ever excites a fear for the issue of a contest which was to change the condition of man over the civilized globe. The sufferings inflicted on endeavors to vindicate the rights of

> humanity are related with the frigid insensibility with which a monk would have contemplated the victims of an auto da fé.[42]

If this criticism did not refer to John Marshall's *Life of Washington,* ironically, it could also be read as a prophetic complaint about the literary merits that Jefferson's own narratives would fail to gain. Rather "frigid" and "partial" itself, the *Autobiography* would turn out to be particularly bad at creating suspense: "No gloom of events, no lowering of prospects ever excites a fear for the issue of a contest." It is remarkable that Jefferson had, in theory, highly literary expectations of historiography, convinced as he was that, like a tragedy, it was supposed to excite "fear." In order to stress the aesthetic and emotive requirements of historical writing, Jefferson established a temperature contrast between the "cold" and "frigid" unenlightened monk on the side of the writer and a warm "kindling" of fiery republican sentiments among the "victims of an auto da fé" on the side of his subject matter. In his nationalized philosophical history, Jefferson expected the historian to abstain from "cold indifference" by imitating the "warmness" of his material and letting the events "kindle" virtuous feelings in his own and his reader's bosom.

The simultaneous accusation of Marshall's "partiality" would not have been an accusation, if Jefferson had not had a positive idea of historical "impartiality." His identification of "partiality" with emotional distance, however, not only runs counter to modern expectations of an "impartial" historian but was very difficult to maintain for Jefferson himself. If the contrary were true, historical "impartiality" would be reflected in the historian's intense emotional involvement with his subject matter–a conclusion that Jefferson could only draw when naively equating the Republican version of American history with "truth." This equation did not work in every context, as his awareness of the partisan "libels" among both Federalist and Republican historians in his letter to Adams suggests. As a result, Jefferson's theory of a literary historiography was as precarious as his conviction of its progressive contents. His Americanized philosophical history had to accomplish too many contrary aims at once: teach universal lessons of a "fixed" historical aim, yet create suspense and "fear" of the outcome; initiate an imitative reading process through the "magic" of his style, yet remain "impartial" by staying true to primary documents. It may have been due to these unresolved issues that Jefferson's *Autobiography* and the *Anas* eventually failed to have the moral and political impact their author had envisioned. Philosophical history, as Jefferson had begun to realize in the late eighteenth century, was constantly in danger of being undermined by its own contradictions. As I discuss in the next section, the internal instability of his history of progress therefore had to be accompanied by historical skepticism and methodological innovations not only during the revolutionary hurricane of the late eighteenth century but also during the little "storms" of the nineteenth.[43]

Awake from Slumber

When describing his hopes for a tranquil detachment from political life during his final retirement, Jefferson liked to employ imagery related to the domain of sleep. For instance, he told the Irish-American journalist William Duane in 1810 that he was hoping for an "undisturbed repose," confident that during Madison's administration he might "sleep in security with such watchmen at the helm . . . and hope to slumber on to my last sleep."[44] The sleep metaphor for Jefferson's final years worked not only mythologically (sleep imagined as Death's brother) but also in Lockean time epistemology. In his *Essay Concerning Human Understanding,* Locke had described duration as the result of the "reflection" on "the constant and regular succession of ideas in a waking man."[45] To this he opposed the experience of sleep:

> When that succession of *ideas* ceases, our perception of duration ceases with it; which everyone clearly experiments in himself, whilst he sleeps soundly, whether an hour or a day, a month or a year; of which duration of things, while he sleeps or thinks not, he has no perception of all, but it is quite lost to him; . . . And so I doubt not but it would be to a waking man, if it were possible for him to keep only one *idea* in his mind, without variation and the succession of others; and we see that one who fixes his thoughts very intently on one thing so as to take but little notice of the succession of *ideas* that pass in his mind whilst he is taken up with that earnest contemplation, lets slip out of his account a good part of that duration.[46]

According to Locke, the absence of time perception during a sleep of undetermined length–apparently, even of whole years–would also be possible in "a waking man" if he managed to keep his thoughts "fixed" on a single idea rather than a succession of ideas. To some extent, this passage also describes the temporality of Jefferson's retirement. He hoped that these years would consist in a long Lockean "sleep," in which he would cease to perceive the turmoil of "human events" even while awake.[47] Trying to fix "his thoughts very intently on one thing," he concentrated on the "object . . . fixed in the eye of nations,"[48] the future achievement of the republican millennium. The more immediate goal Jefferson was envisioning in his personal future was, of course, his death. Habitually referring to this topic in his retirement correspondence, he claimed that he expected to "pass willingly to that eternal sleep, which, whether with, or without dreams, awaits us hereafter."[49] Jefferson stoically expected his approaching death, not as a violent rupture, but as a gradual progression from one state of diminished time perception to the next, until time perception would cease entirely.[50]

This gradualist conception of death–implying that the last seventeen years of his life were a last "slumber" preparing an eternal "sleep"–provided Jefferson with several rhetorical advantages. Not only could he plausibly use the "natural" reduction of his time as an argument to fend off undesired correspondents, including historians whose Republican credentials appeared uncertain.[51] He could also employ allusions to his death to create a truly disinterested, almost posthumous voice–as if already speaking from among the dead,[52] to whom temporal concerns no longer mattered. In this quasi-otherworldly realm, Jefferson could feel himself united with the classical authors whom he admired, declaring to "slumber without fear and review in my dreams the visions of antiquity."[53] Having moved beyond the petty concerns of everyday politics, Jefferson imagined the tranquillity of his retirement as his personal "elysium,"[54] where he could freely converse with the ancients while promoting his classical reputation in a Republican version of the past.

He had long contemplated his future role in history, but it was in his final retirement that he began to think of himself mainly as a historical figure, an ancient "remain" that belonged more to the past than to the present. As one of the last surviving signers of the Declaration of Independence, he professed already to feel like one of the "mere monuments of times past, and specimens of antiquity for the observation of the curious."[55] In this respect similar to Chief Logan of *Notes on Virginia,* Jefferson stylized himself both as part of American "antiquity" and as a member of a past generation.[56] According to generational sovereignty, his generation was morally and politically "dead" by the time of his retirement because its majority was physically dead.[57] It had accordingly lost its "powers and rights" in the present–including, as Jefferson found convenient to point out in his own context, its moral responsibility for present politics (especially on the question of slavery). This self-historicization could occasionally result in a relativistic attitude toward his former political positions, expressing a nuanced awareness of historical change.[58] More often, however, it was part of a self-serving myth about a generation of revolutionaries who had always remained true to their republican principles. The self-righteous note in Jefferson's mythmaking was part of a depersonalized voice that was most effective during the deathlike interval of his retirement "slumber."

Yet Jefferson's repose did not long remain undisturbed. Apart from moments of intense personal worries, for example, concerning his increasingly desperate financial situation, Jefferson also "awoke" from the noise of politics during the War of 1812 and especially during the Missouri Crisis.[59] As he told former Massachusetts congressman John Holmes in an oft-quoted letter written in the spring of 1820, the "momentous question" of the Missouri controversy had, "like a fire bell in the night, awakened and filled me with terror."[60] Jefferson considered the recent compromise in Congress, meant to solve the problem of admitting a slaveholding Missouri to the Union, as an existential threat to American progress. As he told Holmes, a "geographical line, coinciding with a marked principle, moral and political, once conceived and held up to the angry passions of men, will never be obliterated." For Jefferson, the "fire bell" of the Missouri Compromise, which had awakened him from his personal slumber, was therefore "at once . . . a knell to the Union." He feared that the line drawn by Congress between future slaveholding and nonslaveholding states threatened to cast American history back in time, to the level of a belligerent European history. In contrast to his former position, in the Ordinance of 1784, of allowing Congress (even under the Articles of Confederation) to ban slavery in the western territories after 1800, he now accepted the temporary extension of slavery into the West. As Jefferson wrote to Adams, the fact that Congress presumed to regulate slavery could cause "another Peloponnesian war" or at least a "servile war" in America.[61]

Jefferson's being awakened from peaceful dreams by "a fire bell in the night" condensed the psychological shock of a violent displacement from one mode of time perception to another. The image suggests that this transition was forced upon him and that he had played no role in the advent of the catastrophe. He did little to reflect on his possible contribution to the present situation during more than three decades of public service in influential positions. Thus, he failed to ask himself, for instance, whether he had contributed to the strong position Congress assumed in regulating slavery in the West. Since he did not see the past as an organically grown continuum, Jefferson tended to have a narrow concept of historical responsibility. Characteristically, he rationalized the dangers of historical contingency by dividing the past into the clear-cut stages of different generations. In this gradual concept of change, "the generation of 1776" could not be held morally responsible for a crisis in 1820, because it had already fulfilled its providential role in world history by achieving independence from Britain. The Missouri Crisis was due exclusively to the failures of a regressive younger generation who should have moved on to a higher moral level but now even tried to reverse American progress. As Jefferson told Holmes,

> I regret that I am now to die in the belief, that the useless sacrifice by the generation of 1776, to acquire self-government and happiness to their country, is to be thrown away by the unwise and unworthy passions of their sons, and that my only consolation is to be, that I live not to weep over it. If they would but dispassionately weigh the blessings they will throw away, against an abstract principle more likely to be effected by union than by scission, they would pause before they would perpetrate this act of suicide on themselves, and of treason against the hopes of the world.[62]

As discussed earlier, the aged Jefferson often showed himself disappointed with "the post-revolutionary youth,"[63] whom he criticized most of all for their dangerous lack of future-orientation. In 1814 he had lamented the absence of the "good son" in the context of slavery, referring to the quasi-parricidal bloodbath during Priam's suicidal combat in Virgil's rendering of the fall of Troy. As early as 1799 he had prepared the view that his own generation could be followed by a generation who might "lift its parricide hands against freedom & science." As he had tactfully explained to a young man at the time, "That danger comes from the generation your cotemporary."[64] In Jefferson's retirement the classical theme of generational decline after a "Heroic age" gained a new relevance and thus continued into the nineteenth century a long historiographical and literary tradition going back to ideas in Polybius's *Histories* and Hesiod's *Works and Days*.[65]

With the Missouri Compromise, Jefferson feared, the unrestrained "passions" of the sons would make the public-minded actions of their fathers "useless." Essentially, as he saw it, the new generation was dangerously mistaken in its attitude toward time. Its members were unable to "pause" and postpone the gratification of their desires for a moment of rational reflection. Hence, they were unable to "weigh dispassionately" the necessity of preserving their political patrimony against what Jefferson chose to describe merely as an "abstract principle" in this context, the problem of "justice" in the slavery question.[66] They acted rashly and did not take the time to reflect that they were becoming criminals–committing "suicide" on the private level and "treason" on the public level. Thus, they were betraying the progressive "hopes of the world" by destroying their own lives, the liberties of all others, and, without a doubt, the happiness of Jefferson. His tone was so emphatic in this case that it is almost possible to draw a parallel to the voice of the Commendatore in Mozart's *Don Giovanni* (1787):[67] the Holmes letter and the closing of the opera both evoke the uncanny image of a ghostlike appearance admonishing an obstinate young generation to repent for its crimes or die in the flames that the "fire bell" had announced. Unlike the Commendatore, of course, Jefferson had not yet died and was therefore more terrified, himself, by his sudden apparition among the "living generation" and its politics. He was "filled with terror" at the sight of a Europeanization of American history that eluded his modes of prediction and control.

This "terror" was reflected in new metaphors for history and for historical time. History became Europeanized stylistically as well: in moments of crisis or depression, Jefferson tended to describe in European terms events in American history that appeared to interrupt its larger progressive thrust. He referred either to European antiquity–such as to "another Peloponnesian war" in America[68]–or to recent European politics. In order to maintain the moral homogeneity in his national narrative of progress, he metaphorically "denaturalized" northern Federalists by speaking of the "Marats, the Dantons and Robespierres of Massachusetts" in the Hartford Convention.[69] During the Missouri Crisis, somewhat inconsistently with his former metaphor (but consistently placing his own position in the golden mean between two extremes), he referred to Federalist slavery restrictionists as the "Holy Alliance."[70] Thus, he tried to identify as "un-American" those whom he wanted to exclude from his vision of American progress. The disadvantage of this technique was, however, that it led him in fact to construct analogies between recent European and American history, implying that the latter was potentially as regressive as the former.

Although Jefferson's doubts about progress did not usually result in more holistic metaphors for history itself (such as Washington Irving's "stream of history"),[71] they brought along new nuances in his language describing time. As European contingency threatened to invade nineteenth-century America, the mechanistic image of a clock "wound up again" lost plausibility. In the crises of his later years, Jefferson tended, rather, to imagine time in terms of an overpowering nature, such as "the dreary wilderness of time before me" or, echoing the ocean metaphor from Shelley's contemporaneous poem "Time," the "ocean of time opening on us."[72] Time stylistically approached the sublime on land and sea in Jefferson's historical imagery, becoming potentially as "unfathomable" and "shoreless," as "treacherous in calm, and terrible in storm," as in Shelley's poem.[73] When the aging Jefferson saw the years go "beyond the flood,"[74] the "fixed" telos of the republican millennium flowed into an unknowable future. As Jefferson exclaimed in a letter to Albert Gallatin about the Missouri Crisis, the abstractions of philosophical history had become useless in this situation: "What next? Conjecture itself is at a loss."[75]

This uncertainty of American progress was accompanied by a change of emphasis in Jefferson's conception of historical reading. It was perhaps no coincidence that the far greater part of the letter to Adams in which he voiced his sorrows about Missouri for the first time was dedicated, not to present politics, but to reflections on the relationship between ancient and modern history. His reference to the present was relatively brief: he simply observed that the Missouri Crisis was different from other political "occurrences which like waves in a storm will pass under the ship" in that it was "a breaker on which we lose the Missouri country by revolt, and what more, God only knows. From the battle of Bunker's hill to the treaty of Paris we never had so ominous a question."[76] Jefferson thus summed up what he regarded as the historical essence of the Missouri Crisis: as it cast American progress back in time to the level of the Revolutionary War, the Missouri controversy was not a controllable "wave," quickly receding "to it's true level again," but an uncontrollable "breaker," whose future impact on American history, in an unfathomable "ocean of time," completely eluded human knowledge.

Perhaps in an urge to escape from the disorder of the present, in the long remainder of the letter Jefferson was concerned with the still greater political chaos of ancient history. He introduced his musings on the relative shortcomings of Roman republicanism, even among his triad of republican heroes, "Cicero, Cato and Brutus," with a reflection on his own reading process: "I have been amusing myself latterly with reading the voluminous letters of Cicero. They certainly breathe the purest effusions of an exalted patriot, while the parricide Caesar is left in odious contrast. When the enthusiasm however kindled by Cicero's pen and principles, subsides into cool reflection, I ask myself What was that government which the virtues of Cicero were so zealous to restore, and the ambition of Caesar to subvert?"[77]

As discussed earlier, Jefferson tended to group Cicero's letters with Roman historians under the heading "Roman history" (implying that his own correspondence, for instance, would be "American history" as well). In referring to his reading of this historical work, Jefferson emphasized a transition from "enthusiasm" to a more sober level of "reflection." Using the temperature metaphors familiar from his literary criticism of Marshall's *Life of Washington,* he described the sensory appeal of historical reading as something "warm" that was able to "breathe" and "kindle." The combination of Cicero's "pen and principles"–his virtue dressed in a literary style that Jefferson greatly admired[78]–induced readers to imitate Cicero's moral judgments on "patriots" and "parricides" (*parricide,* which is somewhat counterintuitive in the context of Caesar, is used here in the sense of a destruction of the political patrimony). However, in contrast to earlier statements, the aged Jefferson was able to move beyond the aesthetic reading mechanism of philosophical history by letting the "warmness" of his imitative enthusiasm cool down to rational activity.

Whereas he criticized the "frigid insensibility" in Marshall's historical writing, Jefferson now welcomed "cool reflection" in historical reading: it equalized the asymmetry of philosophical history between the process of writing understood as rationally meditated "design" and the process of reading imagined as passive consumption. "Cool reflection" on the side of the reader engendered a critical approach to history, enabling Jefferson to distinguish between Cicero's limited historical perspective and the historical events themselves as they might "really" have been. By asking himself and Adams what the government *was* that Cicero had been trying to restore (in a historical reality that was potentially different from its depiction by "Cicero's pen"), Jefferson moved away from celebrating the "magic" of philosophical history, which potentially reduced the reader, as he put it elsewhere, to a "dupe of mere style."[79] For him, a disenchanted process of historical reading in which readers dissolved the spell of stylistic mastery depended on a temporal factor: the time necessary for letting "warm" emotions "cool" down. This time became relevant not only in the reading of a single work but in the sum of all reading in the course of one's life. As Jefferson's retirement writings suggest, he considered more mature readers like himself better prepared than young readers to achieve the necessary critical distance.

This more active concept of historical reading became important in Jefferson's reaction to David Hume's *History of England.* He was alarmed by the wide discrepancy he perceived between the work's brilliant aesthetics and its shocking political content.[80] After his admiration for Hume's stylistic mastery had made him, as he later remembered, "devour" the *History* with "enthusiasm" as a young man,[81] by the time of his presidency he had developed second thoughts about its political credentials. In about the same period that he became concerned about the dangerous influence of Marshall's *Life of Washington,* Jefferson also began to express fear that the "magic" of Hume's stylistic art would "instil" his partisan "heresies" into the minds of "unweary readers," mechanistically converting them to Toryism or to what amounted to the same thing in Jefferson's dualistic worldview, Federalism.[82] As pragmatically as in Marshall's case, Jefferson began to search for a Republican "antidote" to Hume's "poison." During his second term as president, he thought he had found it in John Baxter's *New and Impartial History of England.* This book was, as Jefferson put it, essentially "Hume's history republicanised,"[83] namely, an abridged and unacknowledged copy of Hume's *History* but with a continuation of the narrative and politically correct alterations.

The "Baxter episode" in Jefferson's relationship to Hume has been criticized for its political dogmatism, at best evincing a lapse from, at worst documenting the overall hypocrisy of, Jefferson libertarianism.[84] Yet in the light of his pragmatic attitude toward history, which valued the moral effect on the reader more than the scholarly integrity of the writer, Jefferson's search for a Republican "antidote" to Hume's widely read *History* hardly appears surprising. Considering the decisive role philosophical history had played for Jefferson from early on, it almost seems surprising that he was not even more dogmatic, but ultimately remained ambivalent toward Hume, the "great apostle of Toryism."[85] Although he tried to promote the publication of Baxter's plagiarism in America, Jefferson's later statements suggest that he was uncertain about the merits of the copy compared with Hume's original. In his retirement writings, he claimed that Baxter had "unfortunately" abridged Hume and that Baxter's continuation until the year 1801 was "a mere chronicle, offering nothing profound."[86] By 1825, while consistently condemning Hume's Toryism, Jefferson was able to historicize the *History of England* itself, conceding that its "unfortunate bias may be partly ascribed to the accident of his having written backwards."[87]

Thus, he renegotiated the unequal, potentially "paranoid" relationship in philosophical history between historians actively mastering their medium and their helpless readers. Jefferson's reflective historical reading now enabled him to conclude that Hume had failed in the neoclassical aim to disguise his will by his "art." The content of Hume's *History* was merely the result of its form, and its form was due more to "accident" than to Hume's intentional "design." In this interpretation, the reader gained rational agency at the expense of the writer (at least if, like Jefferson, readers had had time for "research and reflection"). Rather than dogmatically implying that the reading of Hume's work would invariably produce Tories, Jefferson stressed that it would only be dangerous if it were "put into the hands of . . . our young people."[88] Since Hume's *History,* "were it faithful, would be the finest piece of history which has ever been written by man," its reading should not be abandoned altogether but simply had to be postponed: "Hume, with Brodie [a Whig partisan], should be the last histories of England to be read. If first read, Hume makes an English Tory, from whence it is an easy step to American Toryism."[89]

Whether mainly as a result of his personal exchanges with historians or of his own attempts at historical writing, the study of history had clearly gained intellectual prestige for Jefferson. Commenting, for instance, on the prospectus of a historical work in 1812, he praised it for presenting "a vast anatomy of fact and reflection."[90] Unlike his early philosophical history, historical studies now appeared to require a rational activity comparable to the natural sciences. Less dominated than previously by the literary function of imitating what could have happened, history was supposed self-confidently to deal with the contingent "facts" that had actually chanced to happen. This neo-Aristotelian historiography of Jefferson's later years had a significantly higher status. Especially when he compared it with the novel, he now considered the didactic effect of historical writings far superior to that of merely "fictitious" writings.[91] When he used terms like *romance* or *fable* pejoratively to criticize unfaithful works of history, the traditional hierarchy between poetry and historiography that he had celebrated as a young man was almost reversed.[92]

In the days of the Skipwith letter, Jefferson had negatively defined historiography by its "confinement" to "real life." Since he thought that moral examples were less frequent in historical works than in tragedies and novels, he had considered readers "wisely framed to be as warmly interested for a fictitious as for a real personage."[93] By 1818 he no longer regarded truthfulness to "real life" as a "confinement" in education. To the contrary, he claimed that

> a great obstacle to good education is the inordinate passion prevalent for novels, and the time lost in that reading which should be instructively employed. When this poison infects the mind, it destroys its tone and revolts it against wholesome reading. Reason and fact, plain and unadorned, are rejected. Nothing can engage attention unless dressed in the figments of fancy, . . . The result is a bloated imagination, sickly judgment, and disgust towards the real business of life. This mass of trash, however, is not without some distinction; some few modelling their narratives, although fictitious, on the incidents of real life, have been able to make them interesting and useful vehicles of sound morality. . . . For a like reason, too, much poetry should not be indulged.[94]

In this letter on the education of young women the "real," rather than the probable, had become the touchstone for moral instruction. The aged Jefferson thus left behind the literary gradualism of philosophical history, according to which historiography and poetry had differed only by degrees but not in kind. Instead of emphasizing the universal aim of an imitation of the probable, he now stressed the existence of two opposite realms for history and fiction–the realm of "real life" and the realm of "fancy." Since the 1770s, the content of both novel writing and history writing had changed. Early Romantic novels were moving away from the instructive depiction of the "real" or even the probable to the opposite realm of "fancy," whereas the history of the American republic (at least the Republican version) now found in the "real" enough material for positive moral lessons. The reading of novels and poetry, in Jefferson's view, was rarely "instructively employed" time. It was potentially as "poisonous" to young women as the reading of Hume's partisan *History* was to young men. Novels, as Jefferson thought now, could at best achieve their limited didactic success if "modelling their narratives, *although* fictitious, on the incidents of real life," that is, if they overcame what he now understood as their confinement to the realm of fancy and managed to come as close to "real life" as the unfortunate restrictions of the genre allowed.

By the end of his life, Jefferson's theoretical reflections on the study of history had thus changed in two major respects. Firstly, he had developed a more active and critical concept of historical reading (complemented by a potentially more "accidental" concept of historical writing). And secondly, he now distinguished more clearly between different realms for history and poetry. With his emphasis on "reason" in the process of historical reading and the "real" as the subject matter of historical writing, historical studies had become considerably different from his early view of history as an intellectually unchallenging "moral exercise" whose simple imitation mechanism was best suited for young children. This view had at best been practicable in an ordered universe where the imitation of the probable seemed to guarantee stability in the future. Over the years, however, Jefferson had had many occasions to modify his views, when the teleology of American history appeared disrupted either on the level of the historical events themselves, such as during unpredictable "storms" in his political experience, or on the level of the narration of the historical events, when historians such as Hume or Marshall, like immoral novelists, tried to "seduce" the reading public by their deceptive accounts of the past.

In the more optimistic moments of his retirement, Jefferson showed himself convinced of the great didactic potential of a nationalized philosophical history. If properly composed and decently written, it promised to be the finest moral tale of virtuous Whigs and Republicans, better equipped than merely "fictitious" narratives to instruct republican citizens for the future. However, when regressive events and retrograde writers caused the republican millennium to recede into an unknowable future, Jefferson had to find new ways of dealing with historical studies. In order to preserve at least a part of his optimism about American progress, it became important to modernize his concept of historical reading. Emphasizing the reader's rational activity, he began to loosen the "chain of causes and effects" in the reception of historical as opposed to fictional works. He thus developed an idea of historical-critical reading, while fictional reading remained the domain of the passions and the senses. For Jefferson and many of his contemporaries, the reading of history consequently gained prestige at the expense of the "merely" aesthetic reception of fiction. This priority of history reflected the needs of a first generation of republicans, who after founding the political institutions of the republic also founded its culture of historical memory.

Return to the Magic Mountain

When Thomas Jefferson retired from the presidency in the spring of 1809, he declared himself happy to leave behind "duties which the history of a wonderful era made incumbent on those called upon them." He was delighted to "return" to a place he also defined temporally: the scenes of his "birth and early life" in Virginia.[95] As became clear in the following seventeen years, however, this return was less complete than he had imagined.

At home in Albemarle County, Jefferson's involvement with the "history of a wonderful era" did not end; to the contrary, it became one of the most important endeavors of his retirement to shape the image of himself and his period in the eyes of posterity. Along with his written efforts, Jefferson also sought to construct historical legacies in brick and stone. His posthumous reputation materialized impressively in the founding of the University of Virginia, whose architecture and curricula he shaped decisively,[96] and in the construction of his country mansions, most famously of his half-private, half-public Monticello, completed in 1823.[97]

Swarming with visitors, in the early nineteenth century and still more so today, the pleasant mountaintop gardens and architecture of Monticello have shaped Jefferson's image in the eyes of thousands, even millions, of people over the years.[98] As subject matter for the "moral exercise" of his Americanized philosophical history, the material "remains" from his life may thus have become more influential than his historiographical narratives and document collections (the Declaration of Independence, of course, excepted). Ironically, although it is accessible to many citizens in today's democracy, Monticello hardly conveys historical information presenting Jefferson as a great democrat.[99] In the choice of a mountaintop location, for instance, its builder clearly neglected the practical considerations that would have suited his ideal of a modest republican husbandman, giving priority to the aristocratic privilege of building a Palladian villa on an elevated site instead. The superior position of Monticello enabled Jefferson to use, analogously to his paternalistic foresight in time, his extended vision in space to overlook (and literally look down on) his slaves, the citizens of Charlottesville, and, at the end of his life, the students at the University of Virginia and future leaders of the republic.

The seeming paradox of an aristocratic mansion as a democratic symbol can be explained by a phenomenon that Jefferson and his contemporaries liked to describe as "magic": the aesthetic mode of conveying historical information by the senses rather than by reason, a mode in which Jefferson's architecture was ultimately better prepared to succeed than many of his writings. The final construction of Monticello can be seen as the architectural equivalent to an illusionistic work of philosophical history whose appeal rests in part on the duplicity of its aesthetic strategies. Thus, the mansion pretends to be a modest one-story house, while it really contains hidden second and third stories (an aesthetic pretension of modesty inspired also by contemporaneous European castles). It evokes associations to the Pantheon, the Villa Rotunda, and Chiswick House through an externally conspicuous "dome," which largely remains invisible and dysfunctional on the inside of the house. Monticello's most problematic strategy of disguise concerns the slave quarters, whose lower situation made them invisible from the representative level of the house and garden. Jefferson's Venetian blinds, finally, allowed the builder of the illusion to remain unseen while observing the effect of his art on his visitors.

The resulting discrepancies between form and content have contributed to making Monticello an aesthetic object, a largely depragmatized "monument" and museum piece.[100] Since Jefferson's retirement years, it has been more than a normal planter's mansion–so separated from the functions of a common plantation house that its architect witnessed his own transformation into one of the "mere monuments of times past, and specimens of antiquity for the observation of the curious."[101] The aesthetics of Monticello have thus partly achieved independence from the practical morality of its historical uses. Following the logic of philosophical history, it might be said that the aristocratic architect of Monticello has managed, through the aesthetics of a neoclassical style, to "bewitch" and "seduce" many democratic citizens of today's republic, including those, for instance, who have recently advocated the reappearance of Monticello on the five-cent coin.[102] Moving beyond the communication of a particular moral and political subject matter, the "magic" aesthetics of the place have been able to convince them that a Palladian villa, even a slave plantation, is an appropriate symbol for the American democracy of the twenty-first century.

More than he would probably have liked to know, Jefferson's historical reputation has thus proved the product of stylistic techniques that were similar to those of eighteenth-century masters of style whom he despised. Like them, he has turned out to be a "master magician," as one of his eulogists put it in 1826.[103] Jefferson has achieved his conspicuous place in American culture not primarily through the enumeration of his actions in his autobiographical narratives but in large part due to the aesthetics of his neoclassical style–his "felicity of expression" in the Declaration of Independence, his conscientiously republican style of dress, his cultivated way of living, and, most directly accessible to the senses, the delightful neoclassical facade of Monticello.

These charms, meanwhile, have not been able to put a lasting spell on many scholars. One reason for modern readers to be disenchanted with Jefferson may be that the erudition and relative stiltedness of a neoclassical style lost a part of their "magic" appeal with the emergence of a Romantic conception of artistic genius in the nineteenth century–if not in architecture, in any case in literature. But even for those who still admire Jefferson's artistic talents, it has become impossible to neglect the darker facets of his character. A critical reading of many of his writings tends to reveal a disturbing gap between aesthetics and moral content. If Jefferson's character is interpreted along the "paranoid" lines of philosophical history (presupposing the conscious "design" of the writer or historical agent), the craftsmanship of his writing may seem to have been intended to gloss over the contradictions of its content, just as the habitual politeness of his comportment may appear to have been designed to conceal the intrigues of a clever party politician or his cultured style of living supposed to distract himself and others from a sordid underside of racial exploitation. Some historians have accordingly been just as appalled by Jefferson's double-faced "hypocrisy" as Jefferson himself was, for instance, by Hume, whom he likewise regarded as aesthetically admirable but ethically all the more despicable, at once a stylistic master and a regressive "apostle of Toryism." As Jefferson in retrospect rarely appears to have lived up to the lofty tone of his writings, his strong points on the aesthetic side may in the long run merely draw attention to the contrast of his ethical deficiencies. What once enhanced his historical reputation as an attractive national symbol may therefore eventually serve only to accelerate its decline.

Perhaps, however, Jefferson's style may on the contrary soften the demise of his historical reputation, as it will probably continue, together with several of his political theories, to exert a fascination of its own. It is only possible to play off the aesthetics against the ethics of the historical figure Jefferson, after all, if one remains in the intellectual world of philosophical history–expecting to base universal moral lessons and judgments on the homogeneity of form and content in the past and suspecting any deviance from this homogeneity to be the result of the author's malevolent "design." Even Jefferson himself, I have argued in this study, was not always capable of maintaining this perspective. When his last stronghold of philosophical history–his exceptionalist narrative of American progress–threatened to collapse, he had to leave his didacticism behind, to some extent, and develop a critical concept of historical studies in which aesthetics could be emancipated from their subordination to moral ends. Its many visitors can thus look at Monticello today in the same manner as a sentimental Jefferson looked at ruins from the Roman Empire–as a remnant from a period whose ethics have in part to be questioned but whose aesthetics can still be enjoyed.

Jefferson's "return" to Monticello, it can be concluded, was a return to a place that epitomized the lifelong struggle with style and content in his narrative of national progress. At the same time that it functions as an American work of philosophical history, Monticello is more than a deceptive construct. It is also an authentic historical "remain," if considered as the "real" historical residence of Jefferson and his extended family. It contains some of the furniture they used, books they read, and trees they planted. Comparably to the mounds that Jefferson analyzed as the only "monuments" of Indian history, Monticello now contains the graveyards of his slaves and his family, including the material "remains" of Jefferson himself. Hence, it manages to combine the authenticity of Jefferson's antiquarianism with the polished, potentially insincere "design" of his version of an Americanized philosophical history. In its ambiguity it might be called, in analogy to Thomas Mann's classic novel, a "magic mountain." Although explicitly located on a hill rather than a mountain, Monticello shares an important phenomenon with its equivalent in the modernist novel: as on Mann's fictional magic mountain, time at Monticello seems to flow somewhat differently than it does in the rest of the world. Under the partial spell of its aesthetics, it is still possible to experience the order and homogeneity of Jefferson's philosophical history and his hopes for a lasting control over the future. It seems only appropriate on Jefferson's magic mountain that he did not leave the stage effect to his old friend John Adams at "Montezillo" but likewise crowned his retirement construction of history by the supreme form of a history-conscious time management: the perfect timing of his death on the fiftieth anniversary of American independence.

Epilogue

"I leave it, therefore, to time"

Ten days before his death, Jefferson wrote a letter declining an invitation to attend the celebration of the fiftieth anniversary of the Fourth of July at Washington, DC, because he was too ill to make the journey to the capital. Feeble as he was, he did not let the occasion pass without composing an epistolary valediction praising the crucial significance of his generation's contribution to history. He showed himself convinced that the American Revolution would become

> to the world, what I believe it will be, (to some parts sooner, to others later, but finally to all,) the signal of arousing men to burst the chains under which monkish ignorance and superstition had persuaded them to bind themselves, and to assume the blessings and security of self-government. . . . All eyes are opened, or opening, to the rights of man. The general spread of the light of science has already laid open to every view the palpable truth, that the mass of mankind has not been born with saddles on their backs, nor a favored few booted and spurred, ready to ride them legitimately, by the grace of God.[1]

Jefferson's last message is full of references to temporality, illustrating his own and his generation's middle position in the gradual progress of enlightenment and republicanism. He saw his contribution to this progress, not as its culmination, but as the opening "signal" at the turning point between a Kantian self-incurred tutelage in the *ancien régime* and a new time shaped by enlightened citizens of peaceful republics. Parallel to the signal of the

Declaration of Independence, the empirical sciences had brought to "light" the absurdity of the doctrine of divine right. This visual information, however, had yet to be realized by the majority of mankind. Some "eyes" were already "opened," but from a universal point of view most eyes were still merely in the process of "opening." Within his Newtonian gradualism, Jefferson did not have to commit himself as to the exact temporal scope of this "mean time" between the "signal" of the American Revolution fifty years earlier and the universal vision of human rights at some point in the future. To some parts of mankind, he realized, the self-evident truths of the Declaration would become visible rather "later" than "sooner." Yet what mattered for Jefferson, from the perspective of his final days, was his conviction that the telos of universal history itself was fixed and would "finally" become visible "to all."

The self-congratulatory tone of the letter obscures what may have been the most problematic personal legacy of Jefferson's conception of temporality. He chose to see American republicans as the avant-garde of mankind, since they had already been enjoying the "blessings and security of self-government" for half a century. For this argument, he needed to suppress all mention of the continued existence of chattel slavery in America–an institution that was situated, also by his own standards, at an even lower point of progress than the *ancien régime.* Unenlightened Europeans, after all, had only been persuaded "by monkish ignorance and superstition . . . to bind *themselves,*" whereas American slaves were bound by chains that resulted from the violence of a continued state of war, an archaic conflict that Jefferson had recognized as an unjust war in an enlightened world. If American slaves were aroused by the signal of the American Revolution to claim their universal rights, they had to do more than exit a self-incurred tutelage: they would at first have to repeat the violence of their oppressors and materially "burst the chains" of their bondage in a bloody slave rebellion. In his pessimistic moods, as has been discussed, Jefferson was aware of this crucial difference and feared that apocalyptic scenes caused by a devastating mixture between a "servile war" and a civil war would take place in the future. Yet on occasions allowing or requiring a greater amount of optimism–and the fiftieth anniversary of American independence certainly was among these–Jefferson was able to think of the American slaves merely as one ("African") nation among other nations in Europe and South America, who still had to wait, in the "mean time" of the present, until a future moment when they could follow the example of the American Revolution and found their own republic.

Since for Jefferson the abolition of slavery was necessarily characterized by this alternative–it would be accomplished either by the sudden event of a bloody slave rebellion or by the controlled measures of a gradual emancipation and "expatriation" plan–it seemed plausible that leaving the solution of the problem "to time" was the better option, since a long-term development could be expected to be accompanied by a gradual enlightenment of the slaveholders.[2] In Jefferson's optimistic "dreams of the future,"[3] the gradually enlightened slaveholders would abolish slavery step by step, without the catastrophe of a "servile war." As he had put it in an earlier letter,

> The mind of the master is to be apprized by reflection, and strengthened by the energies of conscience, against the obstacles of self interest to an acquiescence in the rights of others; that of the slave is to be prepared by instruction and habit for self-government, and for the honest pursuits of industry and social duty. Both of these courses of preparation require time, and the former must precede the latter. Some progress is sensibly made in it, yet not so much as I had hoped and expected. But it will yield in time to temperate and steady pursuit, to the enlargement of the human mind, and its advancement in science. We are not in a world ungoverned by the laws and the power of a superior agent. Our efforts are in his hand, and directed by it; and he will give them their effect in his own time. Where the disease is most deeply seated, there it will be slowest in eradication. In the northern States it was merely superficial, and easily corrected. In the southern it is incorporated with the whole system, and requires time, patience, and perseverance in the curative process. That it may finally be effected, and its progress hastened, will be the last and fondest prayer of him who now salutes you with respect and consideration.[4]

This letter summarizes manifold aspects of Jefferson's conception of time in his attitude toward the gradual abolition of slavery. Curing the body politic from the "disease" of slavery, first of all, was only possible by the "temperate and steady pursuit" of long-term measures. In a Newtonian universe, the exact timing of the "curative process" could not be predicted, because the restoration of a previous state of health ultimately depended less on active medication than on prayers to a "superior agent" (as was still the case in most physical diseases in Jefferson's medically primitive America). A Newtonian God would solve the problem of slavery, as Jefferson puts it here, "in his own time," that is, in the "absolute, true and mathematical time" that Newton had described as God's emanation. In Jefferson's approximative empiricism, as has been said, this absolute time was an everlasting version of eternity that was still considered to be the attribute of a benevolent creator and therefore not directly accessible to a limited human understanding. The more enlightened human beings became, the better they could imitate a divine foresight. Yet however enlightened, they would never be able to reach God's absolute control over past, present, and future and the complete knowledge of his providential "design" of the creation.

In a world still governed by Newton's Great Clockmaker, Jefferson found empirical evidence of a providential plan of universal regeneration in the fact that the northern American states, with smaller slave populations, had mostly been cured from the disease, and the mid-Atlantic states were at least advancing in the curative process in proportion to their middling slave population. The spatial gradations in the abolition process, proceeding by degrees from north to south, in an ordered universe could be expected to be largely paralleled by temporal gradations: it would only be natural for the southern states, with the largest slave populations, to be the last states to abolish slavery, since the institution had been deeply implanted into their "system." Since the exact moment of abolition was ultimately unknowable from a limited human perspective on God's providential plan, Jefferson was not surprised that the curative process had turned out to take much longer than he had "hoped and expected" and that he had been mistaken in his early beliefs that abolition would occur in his lifetime. Yet on a more long-term basis, he had no doubt that a natural state of health would be restored, resulting in the eventual deliverance of masters and slaves alike in a secularized eternity.

While the metaphorical process of physical healing stressed the restorative dimension of Jefferson's early historical thought, the second theme in this letter–that of a mental improvement concerning masters as well as slaves–emphasized the conception of progress that had become more prominent in his nineteenth-century version of an Americanized philosophical history. The first step of universal education concerned the masters ("the former must precede the latter"), who had to learn to lead "rational lives" in order to consolidate their identity as natural aristocrats in general and to "hasten" the abolition process in particular. As Jefferson put it in the above letter, before the slaves could receive "instruction," the masters' minds had to be transformed by rational "reflection," allowing them firstly to restrain their "self-interest," secondly to instruct, and thirdly to emancipate their slaves. According to Jefferson's aristocratic ideal of disinterestedness, he expected the masters to abstain from calculating the value of their slaves' labor time in a modern economic sense, in the same way that they often refrained from calculating the role of time in the growth of their debts in a modern financial sense. He hoped they would instead promote the progress of the natural sciences, which, combined with their paternalistic foresight in social concerns, would allow them to plan ahead prudently for the benefit of their extended white and black families.

For this argument, Jefferson had to stress that the black slaves were childlike in their inability to plan ahead for themselves. In his eagerness to show that they were in need of a benevolent *paterfamilias,* as they were deficient in the crucial faculty of rational foresight, Jefferson went beyond a mainstream environmentalism in his suggestion that this deficiency was also caused by some factor connected to their African origin rather than merely by their "habits" in slavery. Even in the above letter detailing the "preparation" of the slaves "by instruction and habit for self-government," he made it clear that this instruction was a passive process of reception, leaving it to the masters actively and rationally to "reflect" on the future. Jefferson's "suspicion only"[5] of an innate rational and physical inferiority of his black slaves is especially disturbing today, as it can be seen to have anticipated nineteenth-century forms of racism. In contrast to that of later racists, however, it was essential to Jefferson's worldview that he could never commit himself in the decisive question of nature or nurture. This inability was grounded in his Newtonian conception of time. Within his flexible dualism of different temporalities, he was able to exclude American slaves from a rational mode of time perception. Yet since this dualism also implied an awareness of the general limits of the human time perception and of his own middle position in the universal progress of knowledge, he always had to admit that his exclusions from rational time could at best be provisional.

Jefferson's emphasis on the limits of human rationality also characterized his views on urban aristocrats in Europe and on older as well as younger generations of Anglo-Americans. According to the aged Jefferson, an absence of rational future-orientation had prevented linear progress in Europe and caused–provoked by the figure of Marie Antoinette as the antithesis of a rational temporality–the radicalism of the French Revolution and the unfortunate return of the *ancien régime.* In America, it had been responsible for the absence of antislavery sentiments among the prerevolutionary generations, brought up in the "quiet and monotonous course of colonial life," and it was at the root of the lamentable polarization of the "post-revolutionary youth" into immediatist critics of slavery, on the one hand, and more outspoken defenders of slavery, on the other.[6] Jefferson knew the appeal of a divergent time management very well. Especially as a Shandean traveler in Europe he had indulged in a sentimental time perception by "liberating" himself from the duties of a forward-looking plantation owner. He always remained aware, however, that such present-centered pleasures had to be enjoyed with moderation and a certain amount of ironic distance. They were supposed to function as a necessary balance to his "rational life," yet never for their own sake, as its Romantic alternative.

From Jefferson's point of view at the end of his life, only his own generation had proved capable of finding a temperate form of time management, during the "Heroic age"[7] of the American Revolution. Even a text like the Declaration of Independence, which implied an abrupt, "immediatist" break with the institutions of the *ancien régime,* was dressed in terms that left all the rashness, violence, and radical immediatism to the actions of George III, while the American side was characterized by a gradualist mentality of future-oriented prudence combined with a profound respect for the past. As Jefferson liked to point out later, his revolutionary generation had accordingly "deserved well of mankind for the struggles it has made,"[8] initiating a process that should have been followed by the next generation's abolition of slavery. The first generation to advocate the abolition of the institution, in a middle position between a less enlightened colonial generation in the past and what they hoped would become more enlightened republican generations in the future, the American revolutionaries had been unable to solve the slavery problem themselves. In Jefferson's retirement his most original contribution to the temporal and historical thinking of his period became his main excuse for his eventual inaction on the slavery question. According to his idea of generational sovereignty, he had become so old as to have "overlived" or "outlived" his political and moral generation. At least in principle, or on occasions when it fitted into his argument, his own position in the gradual sequence of generations forbade him to try to influence decisions in present politics that would shape the lives of a "temporal nation" of which he was no longer a part. His major attempt to tame the "fluidity" of change in a changing world thus also provided him with a conveniently underdeveloped sense of historical causality. Very likely it prevented him from feeling full personal responsibility for negative consequences of political actions that had taken place in what he was able to regard merely as the past of a "dead" generation.

As can be concluded, the elastic categories in Jefferson's transitional thinking about change in time offered him versatile modes of explaining his contradictory position in the central moral question of liberty and slavery. With his flexible eighteenth-century dualism of two secular temporalities, always implying some allowance for the unreliability of a limited perception of the temporal human condition, Jefferson could modify his opinions on the best moment of abolition without necessarily having to contradict himself. During the revolutionary "storm" of the late eighteenth century, he could stress that something needed to be done, "& soon done," in the slavery problem if Americans wanted to avoid the catastrophe of becoming "the murderers of our own children" in a suddenly erupting war caused by the archaic violence at the origin of slavery.[9] In his nineteenth-century efforts to consolidate philosophical history and construct a national narrative of American progress, by contrast, Jefferson could allow the solution of the slavery problem to lose urgency and recede into an increasingly remote future. Depending on the shifting priorities in his interpretation of American history, abolition could thus either occur by unpredictable "human events" in a qualifying sentimental time or form part of a long-term "course" in a quantifying rational time. For Jefferson, the most prudent mode of dealing with this alternative was to unite aspects of both temporalities in the golden mean of his gradual emancipation plans.

The gradualism that characterized Jefferson's attitude toward slavery, then, was not an aberration from his worldview but a central part of it. This worldview dramatically lost plausibility after the transcendental turn, which increased the pressure to take for granted a single concept of time. Seen from the perspective of time philosophy, therefore, it was no coincidence that the response to slavery tended to be radicalized in a Romantic culture, when from the late 1820s on opponents as well as proponents of slavery with growing urgency stressed the needs of their unique moment in the present, whether in the context of evangelical Christianity or in more secular contexts. As the aging Jefferson still realized, the next generations were different from his own in that more Americans either favored immediate emancipation measures or ceased to show any interest in abolition and bluntly argued for maintaining the institution. By comparison, Jefferson found it relatively easier to see himself as an opponent of slavery without feeling compelled to abolish it immediately; similarly, he could advocate the temporary continuation of slavery without being forced to find arguments legitimating the institution as such. Depending on the context, he was able to play the universal off against the particular or the particular against the universal–the everlasting eternity of Newton's absolute time against a limited human perspective of it, or vice versa. Since Jefferson's conception of change in time was never fully secularized, he should not be exclusively identified either with what strikes us today as the most heroic part of his thinking (his universalist, seemingly "democratic" faith in the people) or with what might be viewed as its most villainous aspects (his particularist "racism" and exclusion of various groups from American progress). His gradualism was an integral part of his thinking, allowing him to combine both positions and see degrees and stages where later generations would tend to perceive essential antinomies.

Since for Jefferson the particularities of American "newness" were qualified by the universal goal of his republican millennium, he was not as troubled as later thinkers by the modern problem of an entirely open secular future, as described by a young French aristocrat touring the United States five years after Jefferson's death. At the end of his extensive study, Alexis de Tocqueville concluded that American democracy was such a new phenomenon that the predictable philosophical history of the Enlightenment had lost relevance as a mode of explanation: "Looking back century by century to remotest Antiquity, I see nothing that resembles what I see before me. When the past is no longer capable of shedding light on the future, the mind can only proceed in darkness."[10] In the Newtonian world that had still existed a few years earlier, by contrast, it had continued to make sense for Jefferson to illustrate the passing of time by metaphors emphasizing light rather than darkness, understanding it, in a literal sense, by the analogy to movement in space. As he wrote in 1824:

> I am eighty-one years of age, born where I now live, in the first range of mountains in the interior of our country. And I have observed this march of civilization advancing from the sea coast, passing over us like a cloud of light, increasing our knowledge and improving our condition, insomuch as that we are at this time more advanced in civilization here than the seaports were when I was a boy. . . . Barbarism has, in the meantime, been receding before the steady step of amelioration; and will in time, I trust, disappear from the earth.[11]

Even if European and South American nations still had to wait for "rivers of blood" to "run out" until they would all become republics, and even if American slaves still had to fill "the measure of their tears" until the moment of their emancipation and expatriation, Jefferson showed himself convinced that the "steady step" of gradual progress would "in time" eliminate all forms of barbarism. In the "meantime" of the present, Americans had to do their best, in their most western position of a westward progress, to move the universal process of enlightenment forward. In the context of his only book, the *Notes on the State of Virginia,* Jefferson referred to his contribution to the progress of knowledge by similar metaphors from the domain of light and movement. He modestly described his work, not as a timeless intellectual achievement or an ultimate moral judgment, but as "nothing more than the measure of a shadow, never stationary, but lengthening as the sun advances and to be taken anew from hour to hour."[12] In the extended afternoon of what was still a finite secular time given to mankind by a Newtonian Clockmaker, Jefferson realized that his efforts to understand the spatial and temporal "state" of his native culture could only be provisional: they had to be made again and again by every generation.

If it were true that human knowledge moved with the sun from east to west, Thomas Jefferson's position in intellectual history as discussed in this book could be located geographically in the garden on his little mountaintop in the Piedmont of Virginia. From there he liked to look at the lengthening shadows of the Blue Ridge Mountains in his twilight years, imagining the future states in the American West, the future of mankind generally, and (not least importantly, from his perspective) his own historical reputation in this future. Yet Jefferson never actually traveled beyond the tall mountain ranges that limited his westward vision. As a consequence, it may be concluded, he never saw what was on the other side.

Notes

Introduction

1. Thomas Jefferson (hereafter TJ) to Ellen Randolph Coolidge, 14 November 1825, in Jefferson, *Family Letters,* 460-63.

2. Quoted from TJ's draft of the Declaration in Jefferson, "Autobiography," in *Thomas Jefferson: Writings,* 23.

3. Most famously explained in TJ to James Madison, 6 September 1789, in Jefferson, *Papers of Thomas Jefferson,* 15:392-98 (hereafter cited as Jefferson, *Papers*). This letter is analyzed in chapter 5.

4. See, e.g., TJ to Joseph Priestley, 21 March 1801, in *Thomas Jefferson: Writings,* 1085-87. This letter is discussed in chapter 6.

5. See Jefferson, *Notes,* esp. 102 and 282 (query 11). For TJ's views on the "ancient" culture of Native Americans, see chapter 5.

6. For TJ's own opposition between "backward-looking" and "forward-looking," see, e.g., TJ to Elbridge Gerry, 26 January 1799, TJ to William Green Munford, 18 June 1799, and TJ to Joseph Priestley, 27 January 1800, in Jefferson, *Papers,* 30:645-50, 31:126-30, 339-41. TJ's mode of aestheticizing history as *visual* information is explained in chapters 4-6.

7. Faulkner, *Absalom, Absalom!* 51.

8. Locke, *Essay Concerning Human Understanding,* 104 (2.14.2).

9. On Newtonian concepts of time before the "Kantian solution," see Koyré, *Infinite Universe,* esp. 159-89, 221-72; and Wilcox, *Measure of Times Past,* ch. 2, esp. 23-31. On further implications in eighteenth-century natural and civil history, see Arthur O. Lovejoy's classic *Great Chain of Being;* Lepenies, *Das Ende der Naturgeschichte;* Koselleck, "Historia Magistra Vitae"; and Günther and Koselleck, "Geschichte, Historie."

10. On this tendency in both Newtonianism and Newton's own attitude, see Porter, *Enlightenment,* esp. ch. 6.

11. For a similar emphasis, see Valsania, *Limits of Optimism,* esp. the nuanced interpretation of TJ's dualistic thought in a "culture of limits" in ch. 1.

12. TJ's activities as an amateur horologist are discussed in chapter 1.

13. Most famously in the beginning of Goethe's poem "Den Vereinigten Staaten," in *Zahme Xenien* (1827).

14. See chapter 5.

15. Discussed most influentially in Pocock, *Machiavellian Moment,* ch. 15, esp. 538–52.

16. On this "divide" in progressive conceptions of history, and on the reluctance of nineteenth-century American historians to cross it, see Ross, "Historical Consciousness," 911. TJ's proto-historicism is the subject of chapter 5.

17. For a summary of the discussion around this term, see Cogliano, *Reputation and Legacy,* 8.

18. As summarized in an oft-quoted statement by his early biographer James Parton, "If Jefferson was wrong, America is wrong. If America is right, Jefferson was right." Parton, *Life of Thomas Jefferson,* iii.

19. On the development of TJ's historical reputation, see Peterson, *Jefferson Image;* Wood, "Trials and Tribulations"; French and Ayers, "Strange Career"; Onuf, "Scholars' Jefferson"; Bernstein, *Thomas Jefferson,* 191–98; and Cogliano, *Reputation and Legacy.*

20. TJ to James Madison, 6 September 1789, in Jefferson, *Papers,* 15:392–98.

21. Emerson, "Nature," 3.

22. Jefferson, *Notes,* 63 (query 6); for the discussion of Indian "monuments," see 97–100 (query 11). Apart from Logan's speech, the only other Indian monuments mentioned by TJ were their graves (see chapter 5).

23. TJ to John Adams, 30 August 1787 and 25 March 1826, in *Adams-Jefferson Letters,* 196, 614.

24. For a good example, see Pasley, Robertson, and Waldstreicher, *Beyond the Founders.*

25. For the variety of research on TJ, see the bibliographies by Frank Shuffleton, who lists more than 4,800 titles from TJ's death until 1997, about 30 percent of which have been published since 1980. Shuffleton, *Thomas Jefferson.* The ironical result of TJ's efforts to preserve his papers for posterity is discussed in Cogliano, *Reputation and Legacy,* 98–99.

26. There is a vast literature on TJ's attitude toward slavery. On the literature until 2002, see H. Spahn, *Jefferson und die Sklaverei,* 1–8, 150–53. On the literature since 2002 especially, see Helo and Onuf, "Jefferson, Morality"; Nicolaisen, "Thomas Jefferson, Sally Hemings"; Kennedy, *Mr. Jefferson's Lost Cause;* Wills, *Negro President;* Burstein, *Jefferson's Secrets,* 113–88; Cogliano, *Reputation and Legacy,* 170–229; and Onuf, "Domesticating the Captive Nation."

27. Within the usual paradoxes of the Enlightenment, it is also possible to claim that the reverse was true, that slavery was the foundation of Enlightenment culture. It was the foundation of TJ's progressive thought not only in a material sense, as an economic "base," but also on an intellectual level: TJ expected the universal spread of Enlightenment–which, he hoped, would eventually eliminate slavery–to take place through learning processes, especially reading processes, that were themselves based

on forms of dependence amounting, metaphorically, to slavery. For him, historical agents were, to some extent, Aristotelian "imitative animals," bound by a universal "chain of causes and effects" to the imitation of their "experience" in general, and their reading experience in particular. The problem of liberty and necessity in the empirical Enlightenment surfaced in TJ's views on (historical) literature, as discussed in chapters 4 and 6.

28. The problem of race is examined in chapter 2.

29. TJ to Jean Nicolas Démeunier, 26 June 1786, in Jefferson, *Papers,* 10:63. Démeunier was responsible for preparing the articles on the United States and the separate American states in the section "Economie Politique et Diplomatique" of the *Encyclopédie méthodique.* For Démeunier and TJ's collaboration, see the editorial note in Jefferson, *Papers,* 10:3–11.

30. TJ to Jean Nicolas Démeunier, 26 June, 1786, in Jefferson, *Papers,* 10:63.

31. See the famous passage from query 18 of TJ's *Notes,* originally intended for a French audience: "Indeed I tremble for my country when I reflect that God is just: that his justice cannot sleep forever: that . . . a revolution of the wheel of fortune, an exchange of situation, is among possible events: that it may become probable by supernatural interference! The Almighty has no attribute which can take side with us in such a context. . . . We must be contented to hope they will force their way into every one's mind. I think a change already perceptible . . . the way I hope preparing, under the auspices of heaven, for a total emancipation, and that this is disposed, in the order of events, to be with the consent of the masters, rather than by their extirpation." Jefferson, *Notes,* 163.

32. See J. Adams, "Autobiography," 335.

33. For the tradition of bewilderment at TJ's guarded personality among his biographers, going back to his first professional biographer, James Parton, and condensed in the title of Joseph J. Ellis's *American Sphinx,* see Bernstein, *Thomas Jefferson,* xiv.

34. TJ to John Holmes, 22 April 1820, in *Thomas Jefferson: Writings,* 1434.

35. See TJ to John Adams, 27 August 1786: "I inclose you the article 'Etats Unis' of one of the volumes of the Encyclopedie, lately published. . . . Knowing the importance of setting to rights a book so universally diffused and which will go down to late ages, I answered his queries as fully as I was able, went into a great many calculations for him, and offered to give further explanations where necessary." *Adams-Jefferson Letters,* 152.

36. For an exception, the relatively obscure Scotsman George Wallace, who argued for immediate abolition in the eighteenth century, see Davis, "Early Antislavery Radicalism." On the transition from gradualist to immediatist positions in this period, see Newman, *Transformation of American Abolitionism.*

37. The English term *gradualism* originates from the slavery context, introduced as a criticism by advocates of immediate abolition in the 1830s. See *Oxford English Dictionary,* 2nd ed.

38. TJ began to translate Condorcet's *Réflexions sur l'esclavage des nègres* (first published in 1781) during his time in Paris but never finished. Compare Condorcet, *Œuvres,* 7:61–140; and "Jefferson's Notes from Condorcet on Slavery," in Jefferson, *Papers,* 14:494–98.

39. On the gradualist reforms of slavery in the early republic, see Zilversmit, *First*

Emancipation; Davis, *Slavery in the Age of Revolution;* Wiecek, *Sources of Antislavery Constitutionalism;* and Nash, *Race and Revolution.*

40. Jefferson, "A Declaration by the Representatives of the United States of America, in General Congress Assembled," as inserted into his *Autobiography.* See Jefferson, "Autobiography," in *Thomas Jefferson: Writings,* 22.

41. John Adams to Timothy Pickering, 22 August 1822, in Burnett, *Letters of Members,* 1:516.

42. At least by TJ's own account in the *Autobiography,* delegates from the lower South and from the North who wished to continue the slave trade had been responsible for the deletion: "The clause too, reprobating the enslaving the inhabitants of Africa, was struck out in complaisance to South Carolina and Georgia, who had never attempted to restrain the importation of slaves, and who on the contrary still wished to continue it. Our northern brethren also I believe felt a little tender under those censures; for tho' their people have very few slaves themselves yet they have been pretty considerable carriers of them to others." *Thomas Jefferson: Writings,* 18.

43. Jefferson, "Epitaph" (1826): " 'Here was buried / Thomas Jefferson / Author of the Declaration of Independence / of the Statute of Virginia for religious freedom / & Father of the University of Virginia.' because by these, as testimonials that I have lived, I wish most to be remembered." Facsimile, with the design of his tombstone, in ibid., 706–7. On the problem of TJ's original obscurity as mere "penman" of the document and his own suppressed vanity in the matter, see McDonald, "Jefferson's Changing Reputation," esp. 177–78.

44. See Malone, *Jefferson and His Time,* 1:412–14; and Peterson, *Jefferson and the New Nation,* 281–84.

45. TJ to Jean Nicolas Démeunier, 26 June 1786, in Jefferson, *Papers,* 10:63.

46. TJ to James Heaton, 20 May 1826, in *Thomas Jefferson: Writings,* 1516.

47. Jefferson, *Notes,* 161 (query 17): "Besides, the spirit of the times may alter, will alter."

48. See Sobel, *World They Made Together;* and Breen, *Puritans and Adventurers,* which discuss TJ, among other members of the Virginian planter elite, as typical of an enlightened future-orientation (the "rational time" of chapter 1). Bedini, *Statesman of Science,* examines TJ's interests in the technical aspects of time measurement. More recently, TJ's clocks have been discussed together with his fascinating sketch for a "temporal" monopteros in Allen, *Republic in Time,* 54–58.

49. On the influence of the Whig historians on TJ and other founders, see Colbourn, "Jefferson's Use of the Past"; idem, *Lamp of Experience;* Adair, *Fame and the Founding Fathers,* 107–23; Malone, *Jefferson and His Time,* 6:200–212; and Lienesch, *New Order of the Ages,* esp. 119–37.

50. Recently there have been a number of studies on this topic, taking as their point of departure Douglass Adair's analysis of the founders' concern with the highest of five stages of "fame" (that of *conditores imperiorum*) in his essay "Fame and the Founding Fathers" (1965), reprinted in his book of the same title, 3–26. See Ferling, *Setting the World Ablaze,* ch. 7, esp. 277–84; Sloan, "Presidents as Historians"; Freeman, *Affairs of Honor,* 62–74, 274–88; and Cogliano, *Reputation and Legacy,* 19–105.

51. For TJ's relation to the ancients generally, see K. Lehmann, *American Humanist;* and Richard, *Founders and the Classics.* On classical republican conceptions of

time in his writings, see esp. Pocock, *Machiavellian Moment;* McCoy, *Elusive Republic;* and, in modification of McCoy's argument, Lienesch, *New Order of the Ages,* and Allen, *Republic in Time,* 36–41.

52. For this argument in Newton's context, see Wilcox, *Measure of Times Past,* esp. 49: "New technologies and new critical techniques can be added to old ones; they need not replace the earlier ones completely."

53. Bedini, *Statesman of Science,* 143–44.

54. TJ to Henry Lee, 8 May 1825, in *Thomas Jefferson: Writings,* 1501.

Part I: Time

1. The theorem of a "temporalization" of various domains of eighteenth-century culture was initiated by Lovejoy's classic *Great Chain of Being,* esp. 242–87, "The Temporalizing of the Chain of Being," and continued in Koselleck, "Einleitung," and Lepenies, *Das Ende der Naturgeschichte.* For a criticism of this approach, see Seifert, "Verzeitlichung"; Sawilla, "Geschichte"; and Schinkel, "Imagination as a Category of History."

2. Newton, "Scholium," in *Mathematical Principles,* 8. The scholium to the definitions was already included in the first edition, *Philosophiae Naturalis Principia Mathematica,* from 1687.

3. See Koyré, *Infinite Universe,* esp. 159–89, 221–72; Wilcox, *Measure of Times Past,* 16–31; Padgett, *God, Eternity,* 38–55; and Turetzky, *Time,* 71–76.

4. Newton, *Mathematical Principles,* 8.

5. See, e.g., Laplace's answer to Napoleon's question about the role of God in his *Système du monde:* "Sire, je n'ai pas eu besoin de cette hypothèse." Quoted in Koyré, *Infinite Universe,* 276.

6. See Kant, *Kritik der reinen Vernunft,* 74–93.

7. See, e.g., TJ to Peter Carr, 22 June 1792, in which TJ characterized even Lord Kames as "too metaphysical." Jefferson, *Papers,* 24:107–8.

8. Locke, *Essay Concerning Human Understanding,* 104–8 (2.14.2).

9. Cf. Augustinus, *Confessiones/Confessions,* 230 (conf. 11.14): "What then is time? Provided that no one asks me, I know. If I want to explain it further, I do not know."

10. Locke, *Essay Concerning Human Understanding,* 105–6 (2.14.4). See 2.14.6 for Locke's argument that time perception is *not* linked to external motion.

11. Ibid., 108 (2.14.12).

12. Locke's explanations of this problem remained relatively vague. See ibid., 107 (2.14.9).

13. For Locke's conception of time as the epistemological "mirror image of time in Newton's natural philosophy," see Turetzky, *Time,* 76–79.

14. TJ to Maria Cosway, 12 October 1786, in Jefferson, *Papers,* 10:443–54.

15. See Jefferson, *Papers,* editorial note at 453–54.

16. For example, the Heart was responsible for the public virtues of justice and *amor patriae,* criticizing the Head for living the secluded life of a "monk." Meanwhile, the Head was interested in "public utility," as in its occupation with the architecture of public buildings. Ibid., 450–51, 445.

17. Ibid., 449.

18. See Emerson, "Self-Reliance," 157.

19. TJ to Maria Cosway, 12 October 1786, in Jefferson, *Papers,* 10:446.

20. For TJ's use of the term *history* in reference to his personal life, see TJ to Madame de Tessé, 20 March 1787, ibid., 11:226–28; or TJ to Dr. Vine Utley, 21 March 1819, in *Thomas Jefferson: Writings,* 1416–17.

21. TJ to Maria Cosway, 12 October 1786, in Jefferson, *Papers,* 10:446.

22. Cf. Gibbon's oft-quoted view of history as "indeed, little more than the register of the crimes, follies, and misfortunes of mankind." Gibbon, *Decline and Fall,* 1:95.

1. Rational Time

1. See Breen, *Puritans and Adventurers,* 164–96; Sobel, *World They Made Together,* 48-51, 57–60; and Bedini, *Statesman of Science,* 461–63. The quotation is from TJ to John Adams, 1 August 1816, in *Adams-Jefferson Letters,* 485.

2. Most elaborately discussed in TJ to John Adams, 28 October 1813, in *Adams-Jefferson Letters,* 387–92.

3. Although Defoe's *Robinson Crusoe* was one of the bestselling novels in America in the second half of the eighteenth century, it was not mentioned in TJ's (preserved) library catalogs. See Sowerby, *Catalogue,* 4:445.

4. Defoe, *Robinson Crusoe,* 64, 69, 72.

5. Ibid., 206.

6. Hesiod described the link between time perception, (agri-)culture, and the master-slave relationship in passages in *Works and Days* such as the following: "Mark the days which come from Zeus, duly telling your slaves of them, and that the thirtieth day of the month is best for one to look over the work and to deal out supplies. For these are days which come from Zeus the allwise, when men discern right." Lines 764–69, pp. 58–59 in Evelyn-White's translation.

7. See Sobel, *World They Made Together,* 66, 54. Sobel's interpretation of TJ's views as a reaction against a mentality shaped by slavery is suggestive, although it does not take into account TJ's own inclination toward a slower temporality, discussed in chapter 3.

8. For a discussion of the division of labor in the clock-making industry in the seventeenth and eighteenth centuries, see Thompson, "Time, Work-Discipline."

9. For a comparison to the Chinese fascination with European clocks, see Landes, *Revolution in Time,* 17–52.

10. Jefferson, *Plan for Establishing Uniformity in the Coinage, Weights, and Measures of the United States,* in *Thomas Jefferson: Writings,* 393–415, quotations on 394–95.

11. Landes, *Revolution in Time,* ch. 6; Wendorff, *Zeit und Kultur,* 246–48, 266–74; Whitrow, *Time in History,* 99–114, 139–51.

12. O'Malley, *Keeping Watch,* 56–57.

13. Landes, *Revolution in Time,* 114–31.

14. Koyré, *Infinite Universe,* 273–76.

15. On this symbol in nineteenth- and twentieth-century scholarship from Friedrich Engels to Jacques Le Goff, see Dohrn-van Rossum, "Zeit der Kirche."

16. On King George as a "devoted amateur horologist," see Landes, *Revolution in Time,* 171–72.

17. Bedini, *Statesman of Science,* 200, 502 (doc. 4), 147, 296.

18. Jefferson, *Notes,* 64, 276n102. TJ patriotically thought of renaming the orrery a "Ryttenhouse." Bedini, *Statesman of Science,* 415, 508–9 (doc. 11), 78–79, 143–44.

19. Bedini, *Statesman of Science,* 244–46, 327–29, 506 (doc. 8); Sobel, *World They Made Together,* 57–58.

20. Bedini, *Statesman of Science,* 506 (doc. 8).

21. Ibid., 327, 516 (doc. 19).

22. Ibid., 329.

23. Ibid., 462–63, 512 (doc. 15, "Instructions for the Government of the Artist in Making the Clock for the University of Virginia").

24. TJ designed a special ladder for winding the Great Clock. Bedini, *Statesman of Science,* 328.

25. "Memoirs of a Monticello slave as dictated to Charles Campbell by Isaac," in Bear, *Jefferson at Monticello,* 13.

26. See Sterne, *Tristram Shandy,* esp. 5–6, 8–9.

27. For the technological context of early American clock time, "true" time, and mean time, see O'Malley, *Keeping Watch,* esp. 1–9.

28. For the emergence of clock time since the Middle Ages, see Le Goff, *Pour un autre Moyen Age;* and Dohrn-van Rossum, *History of the Hour.* On the significance of clock time in the American South, see M. Smith, *Mastered by the Clock.* To some extent, TJ's rational time can be read as *avant la lettre* evidence of Smith's thesis (at 4–5) that in the antebellum period clock time helped solve what Eugene Genovese analyzed as the "slaveholders' dilemma," allowing slaveholders to reconcile their contradictory claims to both modernity and a "naturally" grounded social hierarchy. Genovese, *Slaveholders' Dilemma,* esp. 10–45.

29. According to Weber, Christian time-consciousness gained a new dimension in Protestant cultures, where the waste of time became the "first and, in principle, gravest of all sins." Weber, "Protestantische Ethik," 167, my translation.

30. For the contrast of a more prescriptive time in traditional Christianity, see, e.g., the "calendar" and "almanack" attached to the original, 1611 version of the King James Bible, in *Holy Bible: An Exact Reprint.* The 1791 and 1799 editions were used by TJ.

31. See Stein, *Worlds of Thomas Jefferson,* 374–75.

32. For this argument, see Gurevich, "Time as a Problem in Cultural History."

33. See Whitrow, *Time in History,* 157–69; and Landes, *Revolution in Time,* 17–52.

34. The accounts of Webster, Bacon, and Randolph are reprinted in Rosenberger, *Jefferson Reader,* 89–90, 62–66, 67–73.

35. See, e.g., TJ to Thomas Jefferson Randolph, 30 December 1809, in Jefferson, *Family Letters,* 394–95; TJ to John Adams, 21 January 1812, in *Adams-Jefferson Letters,* 292; and TJ to Dr. Vine Utley, 21 March 1819, in *Thomas Jefferson: Writings,* 1416–17.

36. See Jefferson, *Garden and Farm Books;* and *Jefferson's Memorandum Books.*

37. Martin, *Thomas Jefferson, Scientist,* 131–47.

38. *Jefferson's Memorandum Books,* 1:432.

39. Jefferson, *Notes,* 73–81.

40. Bedini, *Statesman of Science,* 118; Jefferson, *Notes,* 278 (ed. n. 1); *Jefferson's Memorandum Books,* 1:771 (ed. n. 2).

41. See the editors' introduction to *Jefferson's Memorandum Books,* 1:xxxvii–xl.

42. Jennings, "Consequences of Prediction," 1134.

43. TJ to Mary Jefferson, 13 June 1790 and 1 March 1791, in Jefferson, *Family Letters,* 58–59, 74.

44. Jefferson, *Notes,* 73.

45. See TJ to Marquis de Chastellux, 2 September 1785, in Jefferson, *Papers,* 8:467–70.

46. Sloan, *Principle and Interest;* introduction to *Jefferson's Memorandum Books,* 1:xviii–xxi.

47. As TJ explained to his grandson, "But where there are no bidders, property, however great, is no resource for the payment of debts." TJ to Thomas Jefferson Randolph, 8 February 1826, in Jefferson, *Family Letters,* 469.

48. See, e.g., TJ to Edward Coles, 25 August 1814, in *Thomas Jefferson: Writings,* 1344–46. On this question, see Stanton, "Those Who Labor for My Happiness."

49. See "Diary of work on the road from the Park branch to Milton. 8 feet wide," in TJ's farm book from December 1795, in Jefferson, *Garden and Farm Books,* 315–16. TJ calculated the costs of the road by a chart containing the dates and time of work (in days and half-days), the number of workers per day ("days work"), and "yards in that time," from which he calculated the average "yds per day for each hand," the total "days work," and the total average of "yds per day." On "dayworks" in a preindustrial context, see Thompson, "Time, Work-Discipline," 395.

50. Jefferson, *Notes,* 143. Cf. TJ to Richard Price, 7 August 1785, in Jefferson, *Papers,* 8:357; and TJ to Edward Coles, 25 August 1814, in *Thomas Jefferson: Writings,* 1345.

51. For his continuing opposition between "interest" and "morality," see TJ to Marquis de Chastellux, 2 September 1785, in Jefferson, *Papers,* 8:467–70; TJ to William A. Burwell, 28 January 1805, in Jefferson, *Works,* 10:126; and TJ to Thomas Law, 13 June 1814, in *Thomas Jefferson: Writings,* 1336. On contrasting redefinitions of this relationship in TJ's intellectual culture, see Hirschman, *Passions and the Interests.*

52. For TJ's own assessment that his debts were due to "my own unskillful management, and devoting my time to the service of my country," see TJ to Thomas Jefferson Randolph, 8 February 1826, in Jefferson, *Family Letters,* 469.

53. Sloan, *Principle and Interest,* 29–31.

54. *Oxford English Dictionary,* 2nd ed.

55. Weber, "Protestantische Ethik," esp. 111–13; Riesman, *Lonely Crowd,* 13–17; Elias, *Prozeß der Zivilisation,* 312–41.

56. TJ to Thomas Jefferson Randolph, 30 December 1809: "For it is only by a methodical distribution of our hours, and a rigorous, inflexible observance of it that any steady progress can be made." Jefferson, *Family Letters,* 395.

57. Riesman, *Lonely Crowd,* 15.

58. Weber, "Protestantische Ethik," 114–16, my translation.

59. See, e.g., TJ to Martha Jefferson, 15 January 1784, in Jefferson, *Family Letters,* 23.

60. See TJ to Martha Jefferson, 21 May 1787 and 28 November 1783, and TJ to Thomas Jefferson Randolph, 30 December 1809, ibid., 41, 19, 395.

61. TJ to Martha Jefferson, 28 November 1783, ibid., 19; TJ to Mary Jefferson, 1 April 1790, ibid., 52; TJ to Francis Wayles Eppes, 9 April 1822, ibid., 443; TJ to Peter

Carr, 19 August 1785, in Jefferson, *Papers,* 8:405–8; TJ to Thomas Mann Randolph Jr., 27 August 1786, ibid., 10:305–9. Cf. Franklin, *Autobiography and Other Writings,* 81.

62. In the tradition of Xenophon and other ancient writers, TJ fundamentally distinguished between farming and more specialized work. A profession was ideally supposed to function only as a supplement to farming. As he wrote to Thomas Jefferson Randolph on 3 January 1809, "I do not know whether your father intends you for a profession, or to be a farmer. This last is the most honourable and happy of all. But farmers as well as professional men are apt to live beyond their income. . . . Under this possibility it is provident and comfortable to possess some resource within ourselves some means by which we can get a living." Jefferson, *Family Letters,* 376. For Franklin's completely different understanding of professionalism, see Franklin's *Autobiography* and his *Way to Wealth,* in Franklin, *Autobiography and Other Writings,* 73, 186–87.

63. TJ to Thomas Jefferson Randolph, 30 December 1809, in Jefferson, *Family Letters,* 394–95.

64. TJ to John Adams, 11 January 1817, in *Adams-Jefferson Letters,* 505; TJ to Francis Wayles Eppes, 6 October 1820, in Jefferson, *Family Letters,* 434.

65. TJ to James Bowdoin, 10 July 1806, in *Writings of Thomas Jefferson,* 11:118.

66. Hamilton W. Pierson, "Jefferson at Monticello: The Private Life of Thomas Jefferson," in Bear, *Jefferson at Monticello,* 84, my emphasis.

67. TJ to Martha Jefferson, 28 March 1787, in Jefferson, *Family Letters,* 34.

68. Cf. TJ to Peter Carr, 19 August 1785, in Jefferson, *Papers,* 8:405–8; and TJ to Thomas Jefferson Randolph, 24 November 1808, in Jefferson, *Family Letters,* 363.

69. See, e.g., Franklin's preface to the 1758 edition of *Poor Richard Improved* for his sayings about the use of time, laziness, and leisure. In many of these Franklin goes further than TJ in promoting wealth and avoiding poverty, e.g., "Laziness travels so slowly that poverty soon overtakes him," "Early to bed, and early to rise, makes a man healthy, wealthy, and wise," "If we are industrious, we shall never starve," "Industry pays debts, while despair increaseth them," and so on. Franklin, *Autobiography and Other Writings,* 186–88.

70. For the classical definition of specialized labor measured in time, making up the "ultimate and real standard by which the value of all commodities can at all times and places be estimated and compared," and its time-saving potential, i.e., "the saving of the time which is commonly lost in passing from one species of work to another," see A. Smith, *Wealth of Nations,* 1:136, 112.

71. See, e.g., TJ to Peter Carr, 19 August 1785, in Jefferson, *Papers,* 8:405–8; and Franklin, *Way to Wealth,* in Franklin, *Autobiography and Other Writings,* 186.

72. TJ to Martha Jefferson, 28 March 1787, and TJ to Francis Wayles Eppes, 6 October 1820, in Jefferson, *Family Letters,* 35–36, 434.

73. TJ to Peter Carr, 19 August 1785, in Jefferson, *Papers,* 8:405–8; TJ to Martha Jefferson, 5 May 1787, in Jefferson, *Family Letters,* 40; TJ to Thomas Jefferson Randolph, 3 January 1809, ibid., 377; TJ to Francis Wayles Eppes, 6 October 1820, ibid., 434; Franklin, *Way to Wealth,* in Franklin, *Autobiography and Other Writings,* 186.

74. These sentences from Voltaire's notebooks, which would find a distant echo in the modernist "recherche du temps perdu," are discussed in Blumenberg, *Lebenszeit und Weltzeit,* 218–41.

75. TJ to Peter Carr, 19 August 1785, in Jefferson, *Papers,* 8:405–8.

76. Rifkin, *Time Wars,* ch. 5.

77. Elias, *Prozeß der Zivilisation,* 2:337.

78. See, e.g., TJ to Martha Jefferson, 28 March 1787, in Jefferson, *Family Letters,* 35; TJ to Eliza House Trist, 15 December 1786, in Jefferson, *Papers,* 10:600; and TJ to F. A. van der Kemp, 11 January 1816, in Jefferson, *Works,* 12:399–401.

79. TJ to Martha Jefferson, 11 December 1783, in Jefferson, *Family Letters,* 21.

80. Ibid.

81. Ibid. Cf. TJ to Peter Carr, 19 August 1785, in Jefferson, *Papers,* 8:405–8.

82. TJ to Martha Jefferson, 11 December 1783, in Jefferson, *Family Letters,* 21.

83. TJ to Martha Jefferson, 28 November 1783, ibid., 20.

84. In the terms of Max Weber's opposition between Leon Battista Alberti's "economic rationalism" and Franklin's concept of a "calling," Jefferson's attitude would be located between the two, arguably closer to the former than to the latter. Weber, "Protestantische Ethik," 63–83, 38n.

85. "Zur Unterordnung kurzfristiger Regungen unter das Gebot einer gewohnheitsmäßigen Langsicht." Elias, *Prozeß der Zivilisation,* 338, my translation.

86. Aristotle, *Politics* 1.2.1252a30–34, 1.4.1253b35, 1.5.1254a22–23, 1255a1.

87. M. Smith, *Mastered by the Clock,* 7, 63.

2. *Paternal Punctuality*

1. Austen, *Mansfield Park,* 206: "Sir Thomas saw them off in as good time as his own correctly punctual habits required." See also 202 and 205. TJ probably preferred Maria Edgeworth's novels to Austen's. TJ to Nathaniel Burwell, 14 March 1818, in *Thomas Jefferson: Writings,* 1411–14.

2. Austen, *Mansfield Park,* 346, 89, 88.

3. Thompson, "Time, Work-Discipline," 361–70.

4. Austen, *Mansfield Park,* 199.

5. On the social significance of epistemology in the scientific revolution, see Shapin, *Social History of Truth,* 65–125, 355–408.

6. Ibid., 462.

7. Defoe, *Robinson Crusoe,* 206.

8. *Jefferson's Memorandum Books,* 1:577 (1 March 1785): "Pd. for a watch for James 46 f."

9. Slaves' frequent ignorance of their age is discussed most famously in Douglass, *Narrative of the Life,* 12.

10. See Equiano, *Interesting Narrative,* 63. For the context of African American temporalities, see Sobel, *World They Made Together,* 30.

11. Bedini, *Statesman of Science,* 295–97.

12. Bedini, "Thomas Jefferson, Clock Designer," 180.

13. Brown, *Wieland, or The Transformation,* 19–23. On TJ's significance for the novel, see Kindermann, *Man Unknown to Himself,* 190–91.

14. On general problems of temporal alterity since the eighteenth century, see Fabian, *Time and the Other* (concerning anthropological discourse); and Kessel, *Langeweile,* esp. 7–14 (concerning gender relations). TJ's ambivalent tendency to deny Native Americans "coevalness," in Fabian's sense, is discussed in my chapter 5.

15. TJ to John Adams, 5 July 1814, in *Adams-Jefferson Letters,* 434.

16. TJ to Anne Willing Bingham, 7 February 1787, in Jefferson, *Papers,* 11:122–24.

17. Anne Willing Bingham to TJ, 1 June 1787, ibid., 11:392–94.

18. TJ to Anne Willing Bingham, 7 February 1787, ibid., 11:122–23.

19. TJ to John Adams (August 1, 1816). *Adams-Jefferson Letters,* 484.

20. TJ to Anne Willing Bingham, 7 February 1787, in Jefferson, *Papers,* 11:123.

21. Ibid., 123–24.

22. Ibid., 123.

23. On women's rationality in the American Enlightenment, see Kerber, *Women of the Republic;* and Kelleter, *Amerikanische Aufklärung,* 708-66. For my interpretation of TJ's views on gender and politics, see H. Spahn, "Silent Course of Happiness."

24. For TJ's views on women and the domestic sphere, see Lewis, "Blessings of Domestic Society."

25. On TJ's conception of geometric, or proportional, equality, see H. Spahn, *Jefferson und die Sklaverei,* 131–42.

26. See also TJ to Anne Willing Bingham, 11 May 1788), in Jefferson, *Papers,* 13:151–52; TJ to John Banister Jr., 15 October 1785, ibid., 8:635–38; and TJ to Peter Carr, 10 August 1787, ibid., 12:14–19.

27. Jefferson, "Autobiography," in *Thomas Jefferson: Writings,* 92.

28. Burke, *Reflections on the Revolution,* 64–66. Burke's personal praise for Marie Antoinette, of course, did not keep him from assuming a close connection between women and nature: without the "ancient chivalry," he wrote, "a king is but a man, a queen is but a woman; a woman is but an animal, and an animal not of the highest order" (66).

29. TJ to Anne Willing Bingham, 11 May 1788, in Jefferson, *Papers,* 13:151–52.

30. TJ to John Adams, 30 August 1787, in *Adams-Jefferson Letters,* 195.

31. Jefferson, "Autobiography," in *Thomas Jefferson: Writings,* 92.

32. TJ to George Washington, 4 December 1788, in Jefferson, *Papers,* 14:328–32.

33. I use TJ's terminology for "blacks" when I describe his perspective since it was basic to his thinking that he did not consider African Americans to be Americans. See Onuf, "To declare them a free and independant people."

34. TJ to Edward Coles, 25 August 1814, in *Thomas Jefferson: Writings,* 1343–46; TJ to John Adams, 25 March 1826, in *Adams-Jefferson Letters,* 614.

35. Jefferson, *Notes,* 142.

36. Ibid., 138.

37. Ibid., 140.

38. Ibid. TJ's criticism of Wheatley is more derisive than his criticism of Sancho, perhaps because of gender or perhaps because he knew of Sancho's correspondence with Laurence Sterne, which had led to the sympathetic description of a "negro girl" in vol. 9, ch. 6, of *Tristram Shandy.*

39. Ibid., 140–43.

40. Ibid., 141.

41. Ibid., 163.

42. Ibid., 142.

43. Ibid., 139.

44. Sobel, *World They Made Together,* 33-34.

45. Jefferson, *Notes,* 139.

46. For an antebellum example, see Hentz, *Planter's Northern Bride,* an 1854 plantation novel that opens with a reference to the historical slaves at Monticello and their alleged love for TJ. In its fictional presentation of an ideal master-slave relationship, the master thinks "deeply," while the slave sleeps "profoundly" (9–10, 31).

47. Locke, *Essay Concerning Human Understanding,* 105 (2.14.4). This topic is continued in chapter 6.

48. Cf. Jefferson, *Notes,* 139; and TJ to Anne Willing Bingham, 7 February 1787 in Jefferson, *Papers,* 11:122–24.

49. Jefferson, *Notes,* 139.

50. Benjamin Banneker to TJ, 19 August 1791, in Jefferson, *Papers,* 22:49–52.

51. TJ to Benjamin Banneker and TJ to Marquis de Condorcet, both 30 August 1791, ibid., 22:97–98, 98–99.

52. TJ to Marquis de Condorcet, 30 August 1791, ibid., 22:99.

53. TJ to Henri Grégoire, 25 February 1809, in *Thomas Jefferson: Writings,* 1202; TJ to Joel Barlow, 8 October 1809, in Jefferson, *Works,* 11:120–24. These letters are among the most compromising documents for potential charges of racism and hypocrisy, as they reveal that TJ was conscious of two different opinions. While retrospectively it may look more likely that TJ was honest in his essentialist view of race and hypocritical when emphasizing a purely environmentalist attitude, it cannot be disproved that he was not hypocritical when writing to both or neither of the correspondents. As he expressed his disgust for any "fixed opinion" even in the letter to Barlow, his "real" opinion may continually have oscillated between the two positions.

54. TJ to Edward Coles, 25 August 1814, in *Thomas Jefferson: Writings,* 1345. For the comparison to children, see also TJ to Edward Bancroft, 26 January 1788, in Jefferson, *Papers,* 14:492.

55. Aristotle, *Politics* 1.2.1252a30–34, 1.4.1253b35, 1.5.1254a22–23, 1255a1. For the comparison, see H. Spahn, *Jefferson und die Sklaverei,* 27–34.

56. Jefferson, *Notes,* 139.

57. Ibid., 138.

58. See W. Jordan, *White over Black,* esp. 486–502.

59. On TJ's relationship with Hemings, see Gordon-Reed, *Thomas Jefferson and Sally Hemings;* Foster et al., "Jefferson Fathered Slave's Last Child"; Lewis and Onuf, *Sally Hemings and Thomas Jefferson;* D. Jordan et al., *Report of the Research Committee;* Lewis et al., "Forum"; Nicolaisen, "Thomas Jefferson, Sally Hemings"; and Gordon-Reed, *Hemingses of Monticello.*

60. Jefferson, *Notes,* 143. For the eighteenth-century terminology, see W. Jordan, *White over Black,* 486–90; and Conze and Sommer, "Rasse."

61. See, e.g., TJ to Henri Grégoire, 25 February 1809: "Whatever be their degree of talent it is no measure of their rights. Because Sir Isaac Newton was superior to others in understanding, he was not therefore lord of the person or property of others." *Thomas Jefferson: Writings,* 1202.

62. TJ to Benjamin Banneker, 30 August 1791, in Jefferson, *Papers,* 22:97–98.

63. About 1788, TJ began to translate Condorcet's *Réflexions sur l'esclavage des nègres,* possibly induced by such flattering statements as the following in the 1788 edition of the text, which he owned: "En Virginie les hommes les plus éclairés s'occupent avec ardeur et avec constance, des moyens de préparer un affranchissement graduel.

L'esclavage est regardé universellement dans les treize Etats comme un crime de lèse-humanité, comme une tache à la gloire des amis de la liberté" (Condorcet, *Œuvres,* 7:139; In Virginia the most enlightened men are engaged in preparing the means for a gradual emancipation with ardor and constancy. Slavery is universally regarded in the thirteen states as an outrage against humanity, as a stain on the glory of the friends of liberty). Cf. "Jefferson's Notes from Condorcet on Slavery," in Jefferson, *Papers,* 14:494–98.

64. Jefferson, *Notes,* 143.

65. TJ to Governor John Langdon, 5 March 1810, in Jefferson, *Papers: Retirement Series,* 2:274–77.

66. TJ to Madame de Tessé, 8 December 1813, in *Thomas Jefferson: Writings,* 1314–16.

67. TJ to Governor John Langdon, 5 March 1810, in Jefferson, *Papers: Retirement Series,* 2:274–77.

68. Jefferson, *Notes,* 143.

69. Ibid.; TJ to Madame de Tessé, 8 December 1813, in *Thomas Jefferson: Writings,* 1315.

70. Jefferson, *Notes,* 163.

71. For the analogy to natural history in the context of political parties, see TJ to John Adams, 27 June 1813, and John Adams to TJ, 13 July 1813, in *Adams-Jefferson Letters,* 335, 355.

72. TJ to John Banister Jr., 15 October 1785, in Jefferson, *Papers,* 8:635–38.

73. Jefferson, *Notes,* 87, 143.

74. Ibid., 163.

75. This fear was especially prominent in TJ's writings from the 1790s. For the witch metaphor, see, e.g., TJ to John Taylor, 4 June 1798, in Jefferson, *Papers,* 30:387–90.

76. Jefferson, *Notes,* 163.

77. This topos was creatively employed in polemics against the *ancien régime* in general. For its use in antislavery rhetoric, see Seeber, *Anti-Slavery Opinion in France.*

78. For the "wolf by the ears" dilemma, see TJ to John Holmes, 22 April 1820, in *Thomas Jefferson: Writings,* 1434.

79. Jefferson, *Notes,* 163.

80. Ibid., 142. Cf. Montesquieu, *De l'esprit des lois,* bk. 15, and idem, "Mes pensées," in Montesquieu, *Œuvres complètes,* 2:492 and 1:1469, respectively.

81. Jefferson, *Notes,* 138, 162–63.

82. Hegel, *Phänomenologie des Geistes,* 107–37, 247–374. The *Phenomenology of Mind* introduced the master-bondsman dialectics most extensively, but Hegel later also referred to it, in modified terms, in his "Philosophie des subjektiven Geistes," in pt. 3 of his *Enzyklopädie der philosophischen Wissenschaften.* For commentary, see Kojève, *Hegel,* 48–89; and Fetscher, *Hegels Lehre vom Menschen,* esp. 95–127.

83. See, e.g., TJ to John Adams, 14 October 1816: "I am not fond of reading what is merely abstract and unapplied immediately to some useful science." *Adams-Jefferson Letters,* 491.

84. For a Hegelian interpretation of American slavery in TJ's time, see Davis, *Slavery in the Age of Revolution,* 557–64.

85. This sentence, present in the original manuscript of *Notes on Virginia,* was later deleted from the text. Jefferson, *Notes,* 292, ed. n. 2.

86. Ibid., 163.

87. Ibid.

88. To come closest to the historical construction of TJ's argument, query 18 has to be read in the terms of a republican tradition in which virtuous manners in general (rather than industrious behavior in particular) are seen as a precondition for the stability of a republic. Cf. query 19 in Jefferson, *Notes,* 165.

89. See J. Miller, *Wolf by the Ears,* 31–37.

90. On the identification of slavery with debt and on TJ's own debts, see Sloan, "Earth Belongs in Usufruct," 288–91.

91. For his ridicule of the "post-revolutionary youth" and its "self-learning," "self-sufficiency," and "intuition" rather than "information of books" and "knoledge acquired in past ages," see TJ to John Adams, 5 July 1814, in *Adams-Jefferson Letters,* 434.

92. TJ to Edward Coles, 25 August 1814), in *Thomas Jefferson: Writings,* 1343–46.

93. Ibid., 1345. TJ quoted from the *Aeneid* 2.509–11, probably from memory, as it is slightly misquoted and misspelled. Cf. Virgil, *Aeneis,* p. 111. TJ also quoted it, with slightly different mistakes, in a letter to John Adams of 27 June 1813. *Adams-Jefferson Letters,* 337.

94. See TJ to James Madison, 6 September 1789, in Jefferson, *Papers,* 15:392–98, discussed in chapter 5. TJ reiterated his views on generational sovereignty in later life, at about the time of his correspondence with Coles, e.g., in TJ to John Wayles Eppes, 24 June 1813, in *Thomas Jefferson: Writings,* 1280–86. For the generational conflict in the context of slavery, see Onuf, "Every Generation is an 'Independant Nation' "; and in the context of the intercultural emergence of "Southern slow time," Sobel, *World They Made Together,* 20, 54–55, 64–67.

95. In Virgil's rendering of the myth (*Aeneid* 2.519–25), nobody bids Priam to take up Hector's arms. On the contrary, his wife, Hecuba, entreats him not to fight. Virgil, *Aeneis,* p. 111.

96. In the same text, for instance, is the figure of Anchises, who escapes the flames of Troy on the shoulders of his son Aeneas.

97. See, e.g., TJ to Marquis de Chastellux, 7 June 1785, in Jefferson, *Papers,* 8:184–86.

3. Sentimental Time

1. Diderot's novel, written in 1773–75, first fully published in French in 1796, with extracts published in Grimm's *Correspondance littéraire* in 1778–80, is listed in Jefferson, "1783 Catalog of Books" (ca. 1775–1812), in Jefferson, *Papers: Electronic Archive,* 202.

2. Diderot, *Jacques le fataliste,* 70: "Car il ne savait que devenir sans sa montre, sans sa tabatière et sans Jacques: c'étaient les trois grandes ressources de sa vie, qui se passait à prendre du tabac, à regarder l'heure qu'il était, à questionner Jacques, et cela dans toutes les combinaisons. Privé de sa montre, il en était donc réduit à sa tabatière qu'il ouvrait et fermait à chacque minute." The master's watch is so useless, in other

words, that the snuffbox can easily take its place, since the master likewise opens and closes it at every minute.

3. On TJ's penchant for Sterne, see the editor's "Register of Authors" in *Jefferson's Literary Commonplace Book*, 182–84; Burstein and Mowbray, "Jefferson and Sterne"; and Burstein, *Inner Jefferson*, 42–67, 99–111.

4. Sterne, *Tristram Shandy*, 5, 9.

5. TJ's generational thought is discussed in chapter 5.

6. See Hume, *Enquiry concerning Human Understanding*, 108–30 (secs. 4–5), esp. 121–22.

7. Kant, "Mutmasslicher Anfang der Menschengeschichte," 90. On Rousseau's influential opposition to a future-oriented life, see Georges Poulet's classic study *Etudes sur le temps humain*, 1:158–73.

8. See Adam Smith's comment on watches in *Theory of Moral Sentiments*, 210–11: "How many people ruin themselves by laying out money on trinkets of frivolous utility?" TJ was certainly in danger of following this tendency: see, e.g., TJ to James Madison, 8 February 1786, in Jefferson, *Papers*, 9:264–67.

9. Poulet, *Etudes sur le temps humain*, 1:xxix; see also 178–93 and 211–17.

10. The only older letter that has been preserved, to one of his guardians, is also about time, although with a rational rather than a sentimental emphasis: it is about TJ's wish to attend college in order to prevent "the Loss of one fourth of my Time . . . by Company's coming here." TJ to John Harvie, 14 January 1760, in Jefferson, *Papers*, 1:3.

11. TJ to John Page, 25 December 1762, ibid., 1:3–6.

12. TJ's uncharacteristic allusion to the devil might have been due to his reading of Daniel Defoe's *Political History of the Devil.* See Sowerby, *Catalogue*, 4:445.

13. TJ to John Page, 25 December 1762, in Jefferson, *Papers*, 1:3–6.

14. TJ to President Monroe, 27 October 1823, in *Thomas Jefferson: Writings*, 1481. These metaphors are discussed further in chapter 6.

15. In his later years he would recommend the instructive qualities of Coke's common-law tracts.

16. On TJ's legal education and the Burwell romance, see, e.g., Peterson, *Jefferson and the New Nation*, 16–20.

17. See *Jefferson's Literary Commonplace Book*, 100–106, 187n.

18. In the "Head and Heart" letter, the Head criticizes the Heart for this sentimental attitude. TJ to Maria Cosway, 12 October 1786, in Jefferson, *Papers*, 10:443–54.

19. See *Jefferson's Literary Commonplace Book*, 130–31, on verses attributed to Anacreon in the eighteenth century, which TJ commonplaced and later partly quoted in Greek in the composition of his "Epitaph" (1826): "Swift as the wheels that kindling roll, / Our life is hurrying to the goal; / A scanty dust, to feed the wind, / Is all the trace 'twill leave behind."

20. See, e.g., *Jefferson's Literary Commonplace Book*, 103.

21. On TJ's admiration for Ossian, see ibid., 141–45, 150–51, 171–73; McLaughlin, "Jefferson, Poe, and Ossian"; and esp. Manning, *Fragments of Union*, 177–95.

22. TJ to Peter Carr, 10 August 1787, in Jefferson, *Papers*, 12:14–19.

23. TJ to Charles McPherson, 25 February 1773: "These pieces have been and will, I

think, during my life, continue to be to me the sources of daily pleasures. . . . I am not ashamed to own that I think this rude bard of the North the greatest poet that has ever existed. Merely for the pleasure of reading his works, I am become desirous of learning the language in which he sung, and of possessing his songs in their original form." Ibid., 1:96–97.

24. TJ liked to employ this reference to the traditional temperaments in order to differentiate himself from his "gloomy" contemporaries. See the oft-quoted letter to John Adams of 8 April 1816: "My temperament is sanguine. I steer my bark with Hope in the head, leaving Fear astern. My hopes indeed sometimes fail; but not oftener than the forebodings of the gloomy." *Adams-Jefferson Letters,* 467.

25. On the source for the proposed tombstone inscription and TJ's modifications, see *Jefferson's Literary Commonplace Book,* 172.

26. TJ to Charles McPherson, 25 February 1773, in Jefferson, *Papers,* 1:96–97.

27. Ossian also played a role in some of TJ's other friendships, such as with the Marquis de Chastellux on his visit to Monticello in 1782. See *Jefferson's Literary Commonplace Book,* 172–73.

28. For the flirtatious uses of Sterne, see, e.g., letters from TJ to Maria Cosway of 12 October 1786 (alluding to the episode with the Franciscan monk in Calais), 24 April 1788 (alluding to Sterne's references to noses), and 27 July 1788 (alluding to Sterne's figure of Maria), in Jefferson, *Papers,* 10:443–54, 13:103–4, 423–24.

29. Jefferson, "Autobiography," in *Thomas Jefferson: Writings,* 46.

30. Jefferson, *Papers,* 6:196. Cf. *Jefferson's Literary Commonplace Book,* 62, 183–84; and Sterne, *Tristram Shandy,* 511.

31. Burstein and Mowbray, "Jefferson and Sterne," 20.

32. See, e.g., Sterne, *Tristram Shandy,* 397: "Pray captain, . . . is a man never overtaken by *Death* in this passage?"

33. For the allusions to Psalms 78:39, Job 7:9, and Proverbs 31:10, see Sterne, *Works,* 3:659n4.

34. Cf. Sterne, *Tristram Shandy,* 511.

35. The next chapter consists in a single ambivalent sentence admitting the extraordinary character of this reflection on human vanity, supposedly denying the significance of "what the world thinks" about it but drawing attention to this question nevertheless: "Now, for what the world thinks of that ejaculation–I would not give a groat." Sterne, *Tristram Shandy,* 512.

36. Although TJ typically claimed to be epicurean rather than stoic, his retirement correspondence is full of stoic references to his death, such as in a Seneca-inspired letter to John Adams of 1 August 1816: "There is a ripeness of time for death, regarding others as well as ourselves, when it is reasonable we should drop off, and make room for another growth. . . . Perhaps however I might accept of time to read Grimm [*Correspondence littéraire*] before I go." *Adams-Jefferson Letters,* 484.

37. See Burstein, *Inner Jefferson,* 60–63.

38. TJ to Elizabeth Wayles Eppes (Martha's half-sister), 3 October 1782: "This miserable kind of existence is really too burthensome to be borne, and were it not for the infidelity of deserting the sacred charge left me, I could not wish for it's continuance a moment." Jefferson, *Papers,* 6:198.

39. TJ to Marquis de Chastellux, 26 November 1782, ibid., 6:203–4.

40. TJ to Elizabeth Wayles Eppes, 3 October 1782, ibid., 6:198.

41. TJ to Elizabeth House Trist, 18 August 1785, ibid., 8:403-5.

42. On these metaphors, see C. Miller, *Ship of State.*

43. TJ to Samuel Henley, 27 November 1785, in Jefferson, *Papers,* 9:65.

44. The letter to Madame de Tott (discussed below) may have been inspired by Sterne's chapters "The Supper" and "The Grace" in vol. 2 of *Sentimental Journey.* Cf. TJ to Madame de Tott, 5 April 1787, in Jefferson, *Papers,* 11:271; and Sterne, *Sentimental Journey,* 118-20.

45. See TJ's travel journals, in *Thomas Jefferson: Writings,* 623-58.

46. TJ to Madame de Tott, 5 April 1787, in Jefferson, *Papers,* 11:271. This letter was not only carefully composed in a Sterne-affecting style but also scrupulously edited by TJ thirty years later. See ibid., 11:272-73, ed. notes. For TJ's own estimate, nevertheless, that "nature had not formed me for a Continuator of Sterne," see TJ to Maria Cosway, 24 April 1788, ibid., 13:103-4.

47. Cf. TJ to Abigail Adams, November 1786, in *Adams-Jefferson Letters,* 157; and TJ to Thomas Jefferson Smith, 25 February 1825, in *Thomas Jefferson: Writings,* 1499-1500.

48. The analogy between Rousseau's presocial and prelinguistic natural state in *Discours sur l'origine et les fondements de l'inégalité parmi les hommes,* published in 1755, and TJ's isolated "little cell" is confined, of course, to the absence of responsibility and future-orientation.

49. For other instances of praise for a Sterne-inspired simple life, see TJ to Maria Cosway, 1 July 1787, and TJ to Alexander Donald, 7 February 1788, in Jefferson, *Papers,* 11:519-20, 12:570-72. For the political dimension of this ideal, see TJ to Marquis de Lafayette, 11 April 1787, ibid., 11:283-85.

50. TJ to Martha Jefferson, 28 March 1787, in Jefferson, *Family Letters,* 34-36.

51. TJ to Maria Cosway, 24 April 1788, in Jefferson, *Papers,* 13:103-4.

52. TJ to Madame de Tessé, 20 March 1787, ibid., 11:226.

53. TJ to Comte Diodati, 29 March 1807, in *Writings of Thomas Jefferson,* 11:181.

54. TJ to Pierre J. G. Cabanis, 12 July 1803, in *Thomas Jefferson: Writings,* 1135-36.

55. TJ to Madame de Staël, 24 May 1813, ibid., 1271-77.

56. TJ's historical outlook on the respective significance of the American and French Revolutions is discussed in chapters 5 and 6.

57. For TJ's reference to Virginia as a "nation," see Jefferson, *Notes,* 162.

58. Locke, *Essay Concerning Human Understanding,* 45 (2.1.2).

59. Ibid., 84-85 (2.10).

60. Ibid.

61. Hume, *Treatise of Human Nature,* 8-10 (1.1.3), 84-86 (1.3.5).

62. TJ to Martha Jefferson Randolph, 8 February 1798, in Jefferson, *Family Letters,* 155-56.

63. TJ to Martha Jefferson Randolph, 15 January 1792, ibid., 93.

64. TJ to Maria Cosway, 12 October 1786, in Jefferson, *Papers,* 10:443-54.

65. TJ to Marquis de Chastellux, October 1786, ibid., 10:498-99.

66. TJ to Jean Nicolas Démeunier, 26 June 1786, ibid., 10:63.

67. TJ to Peter Carr, 10 August 1787, ibid., 12:14-19.

68. On the downside of recollecting the past, see TJ's consolation to John Adams

of 12 October 1813, after the death of his daughter Abigail Smith Adams: "I have ever found time and silence the only medicine, and these but assuage, they never can suppress, the deep-drawn sigh which recollection for ever brings up, until recollection and life are extinguished together." *Adams-Jefferson Letters,* 386.

69. TJ to Peter Carr, 10 August 1787, and TJ to John Banister Jr., 15 October 1785, in Jefferson, *Papers,* 8:635–38.

70. TJ to Gideon Granger, 9 March 1814, in *Writings of Thomas Jefferson,* 14:112.

71. Locke, *Essay Concerning Human Understanding,* 104–8 (2.14). Cf. Sterne, *Tristram Shandy,* 154–56. In contrast to Walter, Toby insists on the subjectivist implications of Lockean temporality: "Now, whether we observe it or no, continued my father, in every sound man's head, there is a regular succession of ideas of one sort or other, . . . which follow and succeed one another in our minds at certain distances, just like the images in the inside of a lanthorn turned round by the heat of a candle.–I declare, quoth my uncle *Toby,* mine are more like a smoak-jack."

72. See Locke, *Essay Concerning Human Understanding,* 107 (2.14.9). TJ may have confused the "lantern" paragraph with the preceding paragraph (2.14.8), as Locke's lantern does not imply a particular "rapidity" of impressions.

73. TJ to John Adams, 5 July 1814, in *Adams-Jefferson Letters,* 430–31.

74. See, e.g., Poe's tales "A Predicament" and "The Devil in the Belfry" and his poem "The Bells," in *Complete Poems and Stories,* 1:241–48, 248–54, 87–90; and Baudelaire's poem "L'Horloge," in Baudelaire, *Les fleurs du mal,* 122–23.

75. See Poe's poem "Dream-land," in *Complete Poems and Stories,* 1:70–71; and Emerson, "Self-Reliance," 157. For the "temporal reforms" of the transcendentalists, see Guthrie, *Above Time.*

76. Burstein, *Jefferson's Secrets,* esp. 33–41.

77. See, e.g., TJ to Eliza House Trist, 15 December 1786, and TJ to Alexander Donald, 7 February 1788, in Jefferson, *Papers,* 8:599–600, 12:570–72; TJ to Abigail Adams, 22 August 1813, in *Adams-Jefferson Letters,* 366–67; and TJ to Edward Coles, 25 August 1814, in *Thomas Jefferson: Writings,* 1343–46.

78. See, e.g., TJ to Peter Carr, 19 August 1785, in Jefferson, *Papers,* 8:405–8.

79. TJ to Horatio Gates Spafford, 11 May 1819, in *Writings of Thomas Jefferson,* 15:189–91; TJ to Abigail Adams, 11 January 1817, in *Adams-Jefferson Letters,* 504.

80. For the argument that the spatial "great chain" itself was in the process of being temporalized in the eighteenth century, see Lovejoy, *Great Chain of Being,* 242–87.

81. For the classic interpretation of this train of thought, see McCoy, *Elusive Republic.*

82. TJ to Anne Willing Bingham, 7 February 1787, in Jefferson, *Papers,* 11:122–24. For another instance of this mentality, see TJ to William Duane, 28 March 1811: "Things may here and there go a little wrong. It is not in their power to prevent it. But all will be right in the end, though not perhaps by the shortest means." Jefferson, *Works,* 11:189–95.

83. TJ to Dr. Thomas Cooper, 7 October 1814, in *Writings of Thomas Jefferson,* 14:199–202.

84. The crucial link between a middle position in the social hierarchy and the moral postulate of a politics of moderation has a long pedigree in Western thought.

For the origins of this link in archaic and classical Greece, from Solon to Aristotle, see P. Spahn, *Mittelschicht und Polisbildung.* For the Aristotelian contrast between the "relatively best" and the "absolutely best" constitution, see idem, "Aristoteles," 428–33.

85. TJ to Edward Coles, 25 August 1814, in *Thomas Jefferson: Writings,* 1343–46. "But in the mean time," wrote Jefferson, "are you right in abandoning this property, and your country with it?" (1346).

86. Jefferson, *Notes,* 163.

87. See, e.g., TJ to John Holmes, 22 April 1820; TJ to Jared Sparks, 4 February 1824; and TJ to James Heaton, 20 May 1826, in *Thomas Jefferson: Writings,* 1433–35, 1484–87, 1516.

88. According to the anti-immediatist rhetoric of the Declaration, only the king's rashness had accelerated history: "Prudence indeed will dictate that governments long established should not be changed for light & transient causes; and accordingly all experience has shown that mankind are more disposed to suffer while evils are sufferable, than to right themselves by abolishing the forms to which they are accustomed. . . . Future ages will scarcely believe that the hardiness of one man adventured, *within the short compass of twelve years only,* to lay a foundation so broad & so undisguised for tyranny over a people fostered and fixed in principles of freedom." TJ's draft as quoted from Jefferson, "Autobiography," ibid., 19, my emphasis.

For another political context, see Malone's interpretation of TJ's willingness to postpone the moment of giving the newly American residents of the Louisiana Territory full political rights: "In times past he [TJ] had shown himself to be a gradualist who could await with patience as well as confidence the attainment of long-range goals. He may have misinterpreted the particular situation to be sure, and he was by no means incapable of self-deception, but it does not necessarily follow that he forgot or disregarded his principles when he recognized that circumstances alter cases." Malone, *Jefferson and His Time,* 4:349.

Part II: History

1. Sterne, *Tristram Shandy,* 70.

2. See TJ's digitized manuscript in the "1783 Catalog of Books" (ca. 1775–1812), in Jefferson, *Papers: Electronic Archive,* 10–12.

3. For a concise description of this calendrical reform, see Zerubavel, *Seven Day Circle,* 28–35; for the thesis of a "new time" in this context, see Koselleck, *Zeitschichten,* 240–45.

4. TJ's interest in promoting the decimal system in other matters, with some reference to the French reform and Talleyrand's proposal to the National Assembly, is documented by his *Plan for Establishing Uniformity in the Coinage, Weights, and Measures of the United States* (4 July 1790), in Jefferson, *Papers,* 16:602–75. Although the decimal system failed to materialize in the United States except in its currency, TJ was very proud of this text, discussing it until his retirement. See TJ to Dr. Robert Patterson, 10 November 1811, in *Thomas Jefferson: Writings,* 1250–58; and TJ to John Quincy Adams, 1 November 1817, in *Writings of Thomas Jefferson,* 15:144–49.

5. On the Whiggish conception of history in TJ's writings, see Colbourn, "Jefferson's Use of the Past"; idem, *Lamp of Experience,* chs. 2–3; Adair, *Fame and the*

Founding Fathers, 107–23; Malone, *Jefferson and His Time,* 6:200–212; and Lienesch, *New Order of the Ages,* esp. ch. 5.

4. Teaching by Examples

1. As a summary term for the foundations of Jefferson's eclectic approach to history, *philosophical history,* with its extensive connotations, seems most adequate. On this concept, see Trevor-Roper, "Historical Philosophy"; White, *Metahistory,* 48–80; Pocock, *Barbarism and Religion,* 7–25; and O'Brien, *Narratives of Enlightenment,* 1–20. For a compelling analysis of the rival term *conjectural history* in this context, see Onuf and Onuf, *Nations, Markets, and War,* esp. 23–29, 42–48.

2. Bolingbroke, *Use of History,* 5 (letter 2); Sowerby, *Catalogue,* 3:132. For a summary of the history books in TJ's libraries, see Colbourn, *Lamp of Experience,* 217–21.

3. Aristotle, *Poetics,* pp. 34–39, 90–93 (chs. 9 and 23).

4. Hume, *Enquiry concerning Human Understanding,* 150 (8.1.7).

5. Bolingbroke's influence on TJ is evident in his *Literary Commonplace Book* and has often been described, for example, in Koch, *Philosophy of Thomas Jefferson.* For the direct influence of Lord Kames on the young TJ, see Helo and Onuf, "Jefferson, Morality,"; and Jayne, *Jefferson's Declaration of Independence.*

6. See Braudy, *Narrative Form,* 31–90, esp. 33.

7. On TJ's preoccupation with Hume's *History of England,* see Walton, "Hume and Jefferson"; and D. Wilson, "Jefferson vs. Hume."

8. Lovejoy, *Great Chain of Being,* 242–87.

9. Wood, "Conspiracy and the Paranoid Style," 413.

10. On the evolution of the term *revolution,* see Arendt, *On Revolution,* ch. 1.

11. These assumptions, of course, extended back to antiquity, but the conviction of the predictability of "human events" may have been especially strong in the Enlightenment, supported by the progress of the natural sciences. For the Renaissance tradition of *ricorso* and *rivoluzione,* tied to *fortuna* or the "wheel of Fortune," see Pocock, *Machiavellian Moment,* esp. 78–79.

12. *Thomas Jefferson: Writings,* 19; Jefferson, *Notes,* 163. In both of these texts, TJ used no single term for the past, but a plural expression referring to the past as a collection of separate events. For the *Kollektivsingular* emerging in the social and political terminology of TJ's period, see Koselleck, "Einleitung."

13. In a traditional use of the term *history* referring not to the past as such but to the literary genre of historical writing, the metaphor of the meteor could be employed for an "irregular" form of writing, especially by people whom TJ excluded from a rational time perception, such as Ignatius Sancho. Bolingbroke used the image to describe a naturally gifted person who had not received a good historical education. See Jefferson, *Notes,* 140; and Bolingbroke, *Use of History,* 7 (letter 2).

14. As a relic from classical and Renaissance thought, *fortune* occasionally appeared in TJ's writings, though mostly in situations requiring a certain amount of pathos, e.g., in query 18, Jefferson, *Notes,* 163. On the traditional concept of *fortuna* as "the symbol of contingency," understood as the "imperfect experience of the perfection of history," see Pocock, *Machiavellian Moment,* esp. 38–40, 156–58.

15. Hume, *Enquiry concerning Human Understanding,* 152–53 (8.12–14).

16. Ibid., 103 (3.9). According to R. G. Collingwood, it was precisely the "false analogy" in philosophical history between the sciences of nature and the "science of man" that distorted the philosophical historians' approach. Collingwood, *Idea of History,* 82–85.

17. Home, *Historical Law-Tracts* (published 1761), v, 22. For commentary, see W. Lehmann, *Henry Home, Lord Kames,* 183–84. The young TJ commonplaced Kames's *Law-Tracts* extensively. See *Commonplace Book of Jefferson,* 95–135.

18. Hume, *Enquiry concerning Human Understanding,* 150 (8.7).

19. Ibid., 152–53 (8.10, 8.15). On TJ's view of individuality, see TJ to Dr. John Manners, 22 February 1814: "Nature has, in truth, produced units only through all her works. Classes, orders, genera, species, are not of her work. Her creation is of individuals. No two animals are exactly alike; no two plants, nor even two leaves or blades of grass; no two crystallizations." *Thomas Jefferson: Writings,* 1329–33. On his ideas of "uniformity" among human beings, see also query 17 of *Notes on the State of Virginia:* "But is uniformity of opinion desireable? No more than of face and stature." Jefferson, *Notes,* 160.

20. J. Adams, "Defence of the Constitutions," 469.

21. See A. Smith, *Theory of Moral Sentiments,* 129.

22. For TJ's architectural penchant for the octagon, see Jefferson, *Architectural Drawings.*

23. Bolingbroke, *Use of History,* 4 (letter 2).

24. Hume, *Enquiry concerning Human Understanding,* 103 (3.9).

25. Momigliano, *Classical Foundations,* esp. 41–44, 49–53. Karl Lehmann names Thucydides as a possible predecessor for TJ's concept of history as "extended, personal experience." K. Lehmann, *American Humanist,* 74.

26. TJ to William Small, 7 May 1775, in Jefferson, *Papers,* 1:165–66.

27. Jefferson, *Notes,* 121.

28. Ibid., my emphasis.

29. This qualification is discussed in chapter 5.

30. TJ to Major John Cartwright, 5 June 1824, in *Thomas Jefferson: Writings,* 1494; *A Summary View of the Rights of British America,* in Jefferson, *Papers,* 1:126.

31. On TJ's concept of human nature, see C. Miller, *Jefferson and Nature,* esp. 6–7, 56–87. See also Curti, *Human Nature in American Thought,* 71–104.

32. Conrad, "*Summary View* Revisited."

33. As argued most prominently in Pocock, *Machiavellian Moment.*

34. For a concise description of the role of the exemplum in Western culture since antiquity, see Engler and Müller, ""Einleitung."

35. Jefferson, *Notes,* 120–21.

36. For TJ's "literal" neoclassicism in public buildings (as opposed to his more flexible Palladianism in private buildings), see, e.g., TJ to William Buchanan and James Hay, 26 January 1786, in Jefferson, *Papers,* 9:220–22; and Jefferson, "Autobiography," in *Thomas Jefferson: Writings,* 41.

37. TJ regarded even his heroes from the late republic, such as Cicero and Brutus, as morally far inferior to modern republicans. See TJ to John Adams, 10 December 1819, in *Adams-Jefferson Letters,* 549.

38. TJ to William Buchanan and James Hay, 26 January 1786, in Jefferson, *Papers,* 9:220–22. Cf. TJ to James Madison, 8 February 1786 and 20 September 1785, ibid., 9:264–67 and 8:534–36.

39. See, e.g., query 11 of *Notes on the State of Virginia,* in Jefferson, *Notes,* 93; and TJ to James Madison, 30 January 1787, in Jefferson, *Papers,* 11:92–97.

40. TJ to John Adams, 27 June 1813, in *Adams-Jefferson Letters,* 335. Cf. TJ to John Taylor, 4 June 1798, in Jefferson, *Papers,* 30:387–90; TJ to Governor John Langdon, 5 March 1810, in Jefferson, *Papers: Retirement Series,* 2:274–77; and TJ to Marquis de Lafayette, 4 November 1823, in *Writings of Thomas Jefferson,* 15:490.

41. TJ to Edmund Pendleton, 13 August 1776, in Jefferson, *Papers,* 1:491–94; TJ to Elbridge Gerry, 26 January 1799, ibid., 30:645–53; TJ to Major John Cartwright, 5 June 1824, in *Thomas Jefferson: Writings,* 1491. Cf. TJ to John Adams, 15 August 1820, in *Adams-Jefferson Letters,* 567.

42. For the topos of "return" or "restoration" as a continuation of the Renaissance concept of *rinnovazione* as the "restoration of true form to human life" and of a Machiavellian *ridurre ai principii,* see Pocock, *Machiavellian Moment,* esp. 108–9, 358–59. TJ also used this topos in a wider sense in a personal context (see chapter 6). For the political concept of restoration and a cyclical theory of history in the American context, see Lienesch, *New Order of the Ages,* 64–68; and Pocock, *Machiavellian Moment,* 506–52.

43. TJ to Edmund Pendleton, 13 August 1776, and TJ to John Randolph, 25 August 1775, in Jefferson, *Papers,* 1:491–94, 240–43. See also 1:339: in his first draft of the Virginia Constitution, Jefferson sought to "re-establish such antient principles as are friendly to the rights of the people."

44. Originally a "Draft of Instructions to the Virginia Delegates in the Continental Congress," it was published in August 1774 as *A Summary View of the Rights of British America.* Ibid., 1:121–35.

45. Conrad, "*Summary View* Revisited," 268–69.

46. On the sense of crisis engendering the question of a "historical future," see Koselleck, *Kritik und Krise,* esp. 105, 132–57.

47. Jefferson, *Summary View,* in Jefferson, *Papers,* 1:125–28, my emphases.

48. Ibid., 132, 122, 135, my emphasis.

49. For the "quest for an enduring 'union' that would sustain and expand the British Empire" in the *Summary View,* see Onuf, *Jefferson's Empire,* 61–63.

50. On this problem, see Fliegelman, *Prodigals and Pilgrims;* and Wood, *Radicalism of the American Revolution,* 11, 145–68.

51. The beginning and ending of the *Summary View* differ substantially in the way they address the king. Beginning as an instruction to the Virginia deputies, the *Summary View* refers to the king in a distanced and indirect manner in the first paragraph, while he is addressed directly in the polemical last paragraph. From the beginning on, however, the *Summary View* presents itself as a necessary refreshment of the king's knowledge of history: "And in order that these our rights, as well as the invasions of them, may be laid more fully before his majesty, to take a view of them from the origin and first settlement of the countries." Jefferson, *Papers,* 1:121.

52. Ibid., 123–24, 134–35.

53. Ibid., 134.

54. Ibid., 134–35.

55. For example, the word *unremitting* and the clause *among which appears no solitary fact to contradict the uniform tenor of the rest* were eliminated by Congress.

56. Bill no. 79, drafted by TJ in 1778, was passed under a new title by the House in 1785 but eventually died after being referred to the Senate. Jefferson, *Papers*, 2:526–35.

57. See Peden, introduction, xi–xx, esp. xiii.

58. Jefferson, *Notes*, 147.

59. Jefferson, *Papers*, 2:526.

60. Jefferson, *Notes*, 148.

61. Ibid. Both texts are ambiguous as to whether TJ expected a good public education merely to postpone corruption or altogether to avoid it. On the role of temporizing in civic humanism, see Pocock, *Machiavellian Moment*, 206.

62. Jefferson, *Notes*, 148.

63. Ibid., 146–48; *Thomas Jefferson: Writings*, 367.

64. Bolingbroke, for example, distinguished explicitly between experience and history, only to say eventually that the two supplemented each other. Bolingbroke, *Use of History*, 6 (letter 2). On the founders' conception of history as "experience," see K. Lehmann, *American Humanist*, esp. 72–79; Colbourn, *Lamp of Experience;* Adair, *Fame and the Founding Fathers*, 107–23; Malone, *Jefferson and His Time*, 6:200–212; and Lienesch, *New Order of the Ages*, ch. 5.

65. Wood, "Conspiracy and the Paranoid Style."

66. Jefferson, *Notes*, 147. For TJ's thoughts on reason in religion, see Sheridan, introduction.

67. Locke, *Educational Writings*, 171–72, 182, 292–94. Cf. Bolingbroke's classical praise of visual examples in *Use of History*, 5–6 (letter 2).

68. Jefferson, *Notes*, 162. Cf. Aristotle, *Poetics*, pp. 12–19 (ch. 4).

69. Jefferson, *Summary View*, in Jefferson, *Papers*, 1:134.

70. For TJ's later characterizations of George III as "only stupid," "without mind" to the point of being "in a straight waistcoat," see, e.g., TJ to Caesar A. Rodney, 10 February 1810, in *Thomas Jefferson: Writings*, 1217; and TJ to Governor John Langdon, 5 March 1810, in Jefferson, *Papers: Retirement Series*, 2:274–77.

71. TJ to Peter Carr, 10 August 1787, in Jefferson, *Papers*, 12:14–19. This letter is also interesting for the controversial question of how far reason and moral sense were able to act independently from each other: "This sense is submitted in some degree to the guidance of reason; but it is a small stock which is required for this." Cf. TJ to Maria Cosway, 12 October 1786, ibid., 10:443–54; and TJ to Thomas Law, 13 June 1814, in *Thomas Jefferson: Writings*, 1335–39.

72. TJ to Robert Skipwith, 3 August 1771, in Jefferson, *Papers*, 1:76–81.

73. TJ to Thomas Mann Randolph Jr., 27 August 1786, and TJ to Peter Carr, 10 August 1787, ibid., 10:305–9, 12:14–19.

74. TJ to Thomas Mann Randolph Jr., 27 August 1786, ibid., 10:305–9.

75. On the presence of this attitude among Voltaire's rather than Montesquieu's followers in philosophical history, see Trevor-Roper, "Historical Philosophy," 1679–80.

76. TJ to Comte Diodati, 29 March 1807, in *Writings of Thomas Jefferson*, 11:181–

83; TJ to John Norvell, 11 June 1807, in *Thomas Jefferson: Writings,* 1176; TJ to John Adams, 1 August 1816, in *Adams-Jefferson Letters,* 485. This attitude is discussed further in chapter 6.

77. On this "cautionary form of history," see Lienesch, *New Order of the Ages,* 123–25.

78. Jefferson, *Summary View:* "a measure which it is believed his majesty's prudence and understanding would prevent him from imitating at this day." Jefferson, *Papers,* 1:123.

79. Cicero's famous praise of history in rhetoric is in *De orat.* 2.9.36: "Historia vero testis temporum, lux veritatis, vita memoriae, magistra vitae, nuntia vetustatis, qua voce alia nisi oratoris immortalitati commendatur?" (Cicero, *De Oratore in Two Volumes,* 1:224–25; And as History, which bears witness to the passing of the ages, sheds light upon reality, gives life to recollection and guidance to human existence, and brings tidings of ancient days, whose voice, but the orator's, can entrust her to immortality?).

80. TJ to Robert Skipwith, 3 August 1771, in Jefferson, *Papers,* 1:76–78. The argument of this letter shows striking similarities to a work on TJ's reading list for Skipwith, which he continued to recommend until his retirement: Home, *Elements of Criticism,* 33–40 (ch. 2.7). Cf. TJ to Samuel R. Demaree, 4 October 1809, in Jefferson, *Papers: Retirement Series,* 1:575–77.

81. Hume, *Enquiry concerning Human Understanding,* 104 (3.10), 102 (3.5–6), 102–3 (3.7–9).

82. According to Collingwood, it was this lack of historical relativization in Hume's conception of mental processes that prevented him from arriving at a modern historical worldview, although, from a more long-term perspective, his "attack on spiritual substance was the philosophical forerunner of scientific history." Collingwood, *Idea of History,* 81.

83. Hume, *Enquiry concerning Human Understanding,* 106 (3.14–15).

84. In his influential "poetics of history," Hayden White has alluded to a return of an eighteenth-century "Ironic" attitude toward historical factuality (which he describes as a "fictionalization" of history and a tendency to view historical writing as "art") in his own formalist approach to nineteenth-century historiography, claiming to hope, however, thus to prepare the ground for another, "anti-Ironic" perspective. White, *Metahistory,* 48–49, 54–59, 64–69, 433–34.

85. Koselleck, "Historia Magistra Vitae," esp. 201–3.

86. TJ to Robert Skipwith, 3 August 1771, in Jefferson, *Papers,* 1:76–78.

87. Aristotle, *Poetics,* pp. 34–35 (ch. 9).

88. Braudy, *Narrative Form,* 91–212.

89. On the *érudits* and the roots of antiquarianism in antiquity, see Momigliano, *Classical Foundations,* 54–79.

90. Bolingbroke, *Use of History,* 3 (letter 1). Cf. TJ to Robert Skipwith, 3 August 1771, in Jefferson, *Papers,* 1:76.

91. TJ commonplaced entries from Bolingbroke's *Philosophical Works,* e.g., paragraphs 57–58. See *Jefferson's Literary Commonplace Book,* 50, 53.

92. Bolingbroke, *Use of History,* 60 (letter 6).

93. While Bolingbroke clearly preferred Livy to the Old Testament, TJ gave more

consideration to the Bible "because millions believe it." *Jefferson's Literary Commonplace Book,* 53. Cf., in Jefferson, *Papers,* 12:14–19.

94. TJ to Peter Carr, 10 August 1787, in Jefferson, *Papers,* 12:14–19, quotation on 15.

95. TJ to Peter Carr, 19 August 1785, ibid., 8:405–8. See also Hellenbrand, *Unfinished Revolution,* 46.

96. Martha admitted to having difficulties with Livy in ancient print: "*Titus Livius* puts me out of my wits. I can not read a word by myself, and I read of it very seldom with my master." Martha Jefferson to TJ, 25 March 1787, in Jefferson, *Family Letters,* 33–34.

97. See, e.g., TJ to Thomas Mann Randolph Jr., 27 August 1786, in Jefferson, *Papers,* 10:305–9. Jefferson may have drawn on Bolingbroke for the criterion of "cotemporary authors" and "cotemporary materials" for historical "authenticity." Cf. *Jefferson's Literary Commonplace Book,* 50–55 (paragraphs 57–58).

98. For the ancients as models of "fine composition," see, e.g., TJ to Joseph Priestley, 27 January 1800, in Jefferson, *Papers,* 31:339–41.

99. K. Lehmann, *American Humanist,* 73–76, 87–90.

100. TJ to Peter Carr, 19 August 1785, in Jefferson, *Papers,* 8:405–8.

101. This method—or lack of one—of grouping together ancient and modern authors may also be seen in TJ's library catalogs and his earlier reading lists. Cf. the Skipwith list, where "HISTORY. ANTIENT" comprises the Bible, Rollin's *Antient History,* translations of Livy and Sallust, and Bayle's *Dictionary.* TJ to Robert Skipwith, 3 August 1771, in Jefferson, *Papers,* 1:76–78. The 1787 reading list for Carr was more systematic, establishing at least a rough chronological order. TJ to Peter Carr, 10 August 1787, ibid., 12:14–19.

102. See, e.g., TJ to Justice William Johnson, 12 June 1823, in *Thomas Jefferson: Writings,* 1471.

103. See, e.g., "The Anas. 1791–1806," ibid., 673.

104. For the rhetorical use of examples, Bolingbroke cited Seneca's "Homines amplius oculis, quam auribus, credunt: longum iter est per praecepta, breve et efficax per exempla," and Pliny's "Mitius jubetur exemplo." Bolingbroke, *Use of History,* 5–6 (letter 2).

105. For TJ's use of the painting metaphor, see, e.g., TJ to Robert Skipwith, 3 August 1771, in Jefferson, *Papers,* 1:76–78; TJ to John Adams, 25 November 1816 and 11 January 1817, in *Adams-Jefferson Letters,* 498, 506; and TJ to William Wirt, 4 and 29 September 1816, in Jefferson, *Works,* 12:32–34, 36–37n.

106. See Schweizer, *Ut pictura poesis Controversy,* 9–52; and Wellberg, *Lessing's Laokoon.*

107. See, e.g., the beginning of Washington Irving's caricature, incidentally, of Jefferson, or "William the Testy," in his satirical *History of New York* (1809): "These and many similar reflections naturally arose in my mind as I took up my pen to commence the reign of William Kieft: for now the stream of our history, which hitherto has rolled in a tranquil current, is about to depart forever from its peaceful haunts and brawl through many a turbulent and rugged scene." Irving, *History of New York,* 166.

108. The term *design* had complex meanings, ranging from an intentional scheme or plot to the visual plan of a work of art. Apart from the religious tradition of model-

ing a master's "design" on God's design in the Creation, it could also have negative connotations in a political context and a relatively technical meaning for the construction of a narrative. For the first, see Wood, "Conspiracy and the Paranoid Style," esp. 403, 411–20; for the second, see, e.g., Hume's first enquiry: "A production without a design would resemble more the ravings of a madman, than the sober efforts of genius and learning." Hume, *Enquiry concerning Human Understanding,* 102 (3.4–5).

109. TJ to John Adams, 5 July 1814, in *Adams-Jefferson Letters,* 433; TJ to Francis Wayles Eppes, 19 January 1821, in Jefferson, *Family Letters,* 438.

110. Jefferson, *Notes,* 177. On the historical works about Virginia at TJ's disposal, see Greene, "Intellectual Reconstruction of Virginia."

111. TJ to Robert Skipwith, 3 August 1771, in Jefferson, *Papers,* 1:76-78.

112. John Adams to TJ, 16 December 1816, in *Adams-Jefferson Letters,* 502–3.

113. TJ to Horatio Gates Spafford, 17 March 1814, in *Writings of Thomas Jefferson,* 14:120.

114. Edward Gibbon's diary, quoted in Trevor-Roper, "Historical Philosophy," 1678. On Gibbon's ambivalence toward Voltaire as a historian, see Pocock, *Barbarism and Religion,* 153-59.

115. TJ to George Washington Lewis, 25 October 1825, in *Writings of Thomas Jefferson,* 16:126. TJ repeatedly tied "magic" to the original power of language, e.g., in *Notes on the State of Virginia:* "To get rid of the magic supposed to be in the word constitution, let us translate it." Jefferson, *Notes,* 124.

116. Ironically, TJ himself was later called a "master magician" in William Wirt's eulogy of 19 October 1826. *Writings of Thomas Jefferson,* 13:xxiii.

117. TJ to Robert Skipwith, 3 August 1771, in Jefferson, *Papers,* 1:77.

118. Ibid.

119. Home, *Elements of Criticism,* 33–40. On further implications of Kames's emphasis on the visual in TJ's context, see Fliegelman, *Declaring Independence,* 16.

120. Jefferson, *Summary View:* "And in order that these our rights, as well as the invasions of them, may be laid more fully before his majesty, to take a view of them from the origin and first settlement of these countries"; TJ to Robert Skipwith, 3 August 1771: "When any original act of charity or gratitude, for instance, is presented either to our sight or our imagination, we are deeply impressed with its beauty and feel a strong desire in ourselves of doing charitable and grateful acts also. On the contrary, when we see or read of any atrocious deed, we are disgusted with it's deformity, and conceive and abhorence of vice." Jefferson, *Papers,* 1:121, 77. For further instances of a connection between the sense of sight and morality, see, e.g., TJ to Thomas Lomax, 12 March 1799, ibid., 31:77–78; and TJ to John Dickinson, 6 March 1801, in *Thomas Jefferson: Writings,* 1084-85.

121. See, e.g., Jefferson, *Notes,* 128–29; TJ to Marquis de Chastellux, 2 September 1785, in Jefferson, *Papers,* 8:467–70; TJ to Thomas Lomax, 12 March 1799, ibid., 31:77–78; TJ to John Norvell, 14 June 1807, in *Thomas Jefferson: Writings,* 1176–79; and TJ to William Duane, 12 August 1810, ibid., 1227–31. On these metaphors in fictional reading, see Fluck, *Das kulturelle Imaginäre,* 38–39.

122. Jefferson, *Notes,* 162.

123. TJ to Robert Skipwith, 3 August 1771, in Jefferson, *Papers,* 1:77.

124. For TJ's fear that malevolent writers might turn their readers into "instruments," see, e.g., TJ to Thomas Lomax, 12 March 1799, ibid., 31:77–78.

125. Bolingbroke, *Use of History,* 6 (letter 2).

5. Seduction by Example

1. TJ to Marquis de Chastellux, 26 November 1782, in Jefferson, *Papers,* 6:203–4.

2. TJ to Comte Diodati, 3 August 1789, ibid., 15:325–27.

3. "The Anas. 1791–1806," in *Thomas Jefferson: Writings,* 662.

4. See Wood, "Conspiracy and the Paranoid Style," 434–35.

5. TJ to Comte Diodati, 3 August 1789, in Jefferson, *Papers,* 15:325–27.

6. Most notoriously in his "Adam and Eve" letter, TJ to William Short, 3 January 1793, ibid., 25:14–17.

7. On the eighteenth-century debates about the relationship between the French and American Revolutions, with a helpful overview of recent scholarship on the issue, see Ziesche, *Cosmopolitan Patriots.*

8. TJ to James Madison, 9 June 1793, in Jefferson, *Papers,* 26:239–42.

9. Berlin, *Proper Study of Mankind,* 243–68, 326–58. For an alternative to Berlin's approach in this context, see Pocock, *Barbarism and Religion,* 23. Pocock bases his study of Gibbon on Momigliano's triad of narrative, erudition, and philosophy instead of on a movement from "Enlightened naturalism" to "a romantic and largely east European historicism," but he admits that this "movement certainly occurred." For approaching the partial breakdown of philosophical history in the dualistic thought of TJ, whose major reflections on history started considerably later than Gibbon's, I found it helpful to use the larger categories of Berlin's "Counter-Enlightenment" argument (whether of "east," central, southern, or general Euro-American provenance).

10. I use *historicist* in the sense of Friedrich Meinecke's rather than Karl Popper's use. For the relation between eighteenth-century *historism* and philosophical history, see White, *Metahistory,* 48–80; and Megill, "Aesthetic Theory and Historical Consciousness."

11. See TJ to James Madison, 30 January 1787, in Jefferson, *Papers,* 11:92–97.

12. The draft is quoted from Jefferson, "Autobiography," in *Thomas Jefferson: Writings,* 19–20. For the religious overtones of this sentence, see Manning, *Puritan Provincial Vision,* 59.

13. Jefferson, *Notes,* 118–29.

14. TJ's criticism of a dictatorship in Virginia may have been exceptionally emotional for personal reasons as well: the second dictatorship proposal had followed directly on his problematic retirement from the governorship of a war-torn Virginia in June 1781. See ibid., 284n28.

15. "All bad examples are derived from good ones; but when power comes to the ignorant and the less good, the new example is transferred from the worthy and fit to the unworthy and unfit." Ibid., 126. Cf. Bolingbroke's discussion of the dangers of directly translating historical examples, for which he quotes Guicciardini rather than Machiavelli. Bolingbroke, *Use of History,* 18–19 (letter 3).

16. Jefferson, *Notes,* 128–29.

17. Ibid., 129.

18. Ibid., 138, 137–43 (discussed in chapter 2). In the context of slavery, TJ linked the moral inferiority of the ancient Romans and the racial inferiority of African slaves, claiming that the institution of slavery was more humane in modern than in ancient times and thus inverting a widespread eighteenth-century evaluation of ancient and modern slavery (as in book 15 of Montesquieu's *De l'esprit des lois).*

19. TJ to Madame de Tessé, 20 March 1787, in Jefferson, *Papers,* 11:226.

20. For TJ's fashionable flirtation with secondary Dutch masters in the 1780s, see TJ to Maria Cosway, 24 April 1788, in Jefferson, *Papers,* 13:103–4.

21. TJ to William Short, 25 August 1790; TJ to George Washington, 14 August 1787; TJ to the Virginia delegates in Congress, 12 July 1785 (containing TJ's recommendation of a small size for Houdon's equestrian statue of Washington for Congress); TJ to James Madison, 8 February 1786 (on Latin mottoes for Washington's statue); TJ to John Trumbull, 15 February 1789 (ordering "the size of the life" for his copies of paintings of Bacon, Locke, and Newton), in ibid., 17:421–23, 12:36–38, 8:289–99, 9:264–70, 14:561.

22. TJ to Nathaniel Macon, 22 January 1816, in *Writings of Thomas Jefferson,* 14:408–10.

23. See Shackleford, *Jefferson's Travels in Europe,* 23–24.

24. For the radicalized emphasis on the contemporary in David's *Serment du jeu de paume* (1791) and its switch from classical to period costume, see Ritter, "Du sollst deiner eigenen Zeit angehören."

25. TJ to Thomas Mann Randolph Jr., 27 August 1786, in Jefferson, *Papers,* 10:305–9.

26. For the ancient tradition of stressing the historian's reliance on the evidence of reliable witnesses, rather than written evidence, see Momigliano, *Classical Foundations,* esp. 39–44.

27. TJ to George Washington Lewis, 25 October 1825, in *Writings of Thomas Jefferson,* 15:124.

28. On TJ's exchanges with these historians, see Cogliano, *Reputation and Legacy,* ch. 2.

29. TJ to William Gordon, 2 July 1787, in Jefferson, *Papers,* 11:525. On the contemporary histories of the American Revolution, see Shaffer, *Politics of History;* and Cohen, *Revolutionary Histories.*

30. TJ to Jean Nicolas Démeunier, 26 June 1786, in Jefferson, *Papers,* 10:364–83, 61–64.

31. In his "observations" on Démeunier's article, as discussed in the introduction, TJ attempted to excuse the fact that the Virginia assembly had failed to provide a clause for emancipation in its new slave law. TJ to Jean Nicolas Démeunier, 26 June 1786, ibid., 10:61–64.

32. TJ was anything but content, even with the final version of Démeunier's *Encyclopédie* article, which turned out to be more favorable than Soulés's *Histoire.* See TJ to John Adams, 27 August 1786, in *Adams-Jefferson Letters,* 152.

33. TJ to John Adams, 11 July 1786 and 15 June 1813, ibid., 143, 333.

34. TJ to John Adams, 15 August 1820, ibid., 567–69. Adams likewise admitted to being "afflicted with a kind of Pyrrhonism" in history and chronology, as in his letter

to TJ of 16 July 1814, ibid., 434–39. On TJ's antiskeptical reading, see Jayne, *Jefferson's Declaration of Independence,* ch. 5, esp. 92–98.

35. TJ to Marquis de Chastellux, October 1786, in Jefferson, *Papers,* 10:489–90; Jefferson, *Thoughts on English Prosody,* in *Thomas Jefferson: Writings,* 593–622.

36. TJ to Marquis de Chastellux, October 1786, in Jefferson, *Papers,* 10:490.

37. *Thoughts on English Prosody,* in *Thomas Jefferson: Writings,* 594.

38. For instance, as "evidences of the age of the writing," TJ carefully preserved the "orthography, abbreviations, interlineations, and incorrectnesses" in his copy of Thomas Mathew's manuscript account of Bacon's Rebellion. See Sowerby, *Catalogue,* 1:226 (no. 534). For this quotation in the context of TJ's antiquarianism, I thank J. Jefferson Looney for letting me read his paper "Beyond the Reach of Accident."

39. TJ to the editor of the *Journal de Paris,* 29 August 1787, in Jefferson, *Papers,* 12:61–65.

40. Ibid., 12:65, quoted from the review in the editorial note.

41. Ibid., 12:64, as quoted twice by TJ from the review.

42. Jefferson, "Epitaph" (1826), in *Thomas Jefferson: Writings,* 706–7.

43. McDonald, "Jefferson's Changing Reputation," 169–79.

44. TJ to the editor of the *Journal de Paris,* 29 August 1787, in Jefferson, *Papers,* 12:63.

45. TJ loved to point out this detail later, as in a letter to Samuel Adams Wells of 12 May 1819: "It was signed by every member present, except Mr. Dickinson. . . . The convention . . . named a new delegation on the 20th, leaving out Mr. Dickinson, who had refused to sign." *Thomas Jefferson: Writings,* 1419–22.

46. TJ to John Trumbull, 30 August 1787, in Jefferson, *Papers,* 12:69.

47. See their intense correspondence in the summer and fall of 1787 in ibid., 12:60, 69, 139, 206–7, 358–59, 405–6.

48. For a reprint of TJ's (incorrect) floor plan and Trumbull's first sketch of the *Declaration* (with a tall figure in the center) from September 1786, see ibid., 10:178. On TJ's influence on Trumbull, see W. Adams, *Paris Years,* 90–98. On Trumbull's stay at TJ's home in Paris, see Malone, *Jefferson and His Time,* 2:138–41.

49. Jefferson, *Papers,* 15:239–41.

50. TJ to John Paradise, 5 July 1789, in ibid., 15:242, discussed in W. Adams, *Paris Years,* 96–97. Cf. Home, *Elements of Criticism,* 388: "In history painting, the principal figure is placed in the front, and in the best light." It might be argued that Trumbull had diplomatically arranged for TJ and John Adams to share this position.

51. For a concise description, see Schweizer, *Ut pictura poesis Controversy,* esp. 29–106.

52. TJ began to modify his classical painting metaphors for history in order to include temporal change in the static space of his historical "paintings," referring to history not as an accomplished work but as a painter's work in progress, a "sketch" or a "canvas." For the dynamic canvas metaphor for history, see TJ to James Madison, 31 July 1788, in Jefferson, *Papers,* 13:440–44; TJ to Count de Moustier, 17 May 1788, ibid., 13:173–76; TJ to Rabaut de St. Etienne, 3 June 1789, ibid., 15:166–67; and TJ to C. F. de C. Volney, 8 February 1805, in *Thomas Jefferson: Writings,* 1154–58.

53. Home, *Elements of Criticism,* 37.

54. Aristotle, *Poetics,* pp. 34–39, 90–93 (chs. 9 and 23).

55. TJ to Samuel Adams Wells, 23 June 1819, in Jefferson, *Works,* 12:126n. TJ referred to this exchange two years later in his *Autobiography.* See Jefferson, "Autobiography," in *Thomas Jefferson: Writings* 7, 24.

56. Cf. Home, *Elements of Criticism,* 37: "A good historical picture makes a deeper impression than words can."

57. *On the spot* was a typical expression of Jefferson's to support his claim to historical truth. Apart from the letter to the *Journal de Paris* ("I was on the spot, and can relate to you this transaction with precision" [Jefferson, *Papers,* 12:61–64]), see TJ to Richard Price, 8 January 1789, ibid., 14:420–24; or TJ to Samuel Adams Wells, 23 June 1819: "But what I wrote on the paper from which I sent extracts to you, was written on the spot, in the moment, and is true." Jefferson, *Works,* 12:126n.

58. For the differing conceptualizations of virtue in this period, see Schloss, *Die tugendhafte Republik.*

59. Bolingbroke, *Use of History,* 5 (letter 2).

60. TJ to Justice William Johnson, 12 June 1823: "Our opponents are far ahead of us in preparations for placing their cause favorably before posterity. . . . History may distort truth, and will distort it for a time, by the superior efforts at justification of those who are conscious of needing it most." *Thomas Jefferson: Writings,* 1469–71.

61. On this point see TJ to Ebenezer Hazard, 18 February 1791; TJ to George Wythe, 16 January 1796; and TJ to William Green Munford, 18 June 1799, in Jefferson, *Papers,* 19:287–89, 28:583–55, 31:126–30.

62. TJ to John Adams, 10–11 August 1815, in *Adams-Jefferson Letters,* 452.

63. TJ's collection of colonial documents has been digitized in Series 8, Virginia Records, 1606–1737, *Jefferson Papers, 1606–1827.*

64. TJ to George Wythe, 16 January 1796, in Jefferson, *Papers,* 28:583–91.

65. TJ to Robert Skipwith, 3 August 1771, ibid., 1:76–78.

66. In these years, TJ came closest to a historiographical interpretation of American history in his "Notes on Professor Ebeling's Letter of July 30, 1795," advising Christoph Daniel Ebeling, a historian and geographer from Hamburg, on his planned historical work on the American Revolution (published as *Erdbeschreibung und Geschichte von Amerika*). Yet these "notes" contained merely a short narrative passage supplemented by lists of trustworthy writers, newspapers, and histories. See Jefferson, *Papers,* 28:423–27. It was probably not before 1800 that TJ composed a short list of his own contributions to American history: "A Memorandum (Services to My Country)," in *Thomas Jefferson: Writings,* 702–4.

67. On TJ's organization of his own papers, see Cogliano, *Reputation and Legacy,* 74–105.

68. "The Anas. 1791–1806," in *Thomas Jefferson: Writings,* 661.

69. For TJ's use of the term *authentic,* see, e.g., his explanations to the "Anas," ibid.; and TJ to John Adams, 11 January and 8 September 1817, in *Adams-Jefferson Letters,* 506, 520.

70. TJ to John Trumbull, 13 November 1787, in Jefferson, *Papers,* 12:358.

71. See W. Adams, *Paris Years,* 78–122; and Shackelford, *Jefferson's Travels in Europe,* esp. 15–30, 103–8, 164–65.

72. TJ to John Trumbull, 15 February 1789, in Jefferson, *Papers,* 14:561.

73. On the importance of TJ's European portraits for his historical reputation, see G. Wilson, "Behold me at length."

74. TJ to James Madison, 20 September 1785, in Jefferson, *Papers,* 8:534–37.

75. Cf. a similar use of the term *national* in Jefferson, *Notes,* 162.

76. TJ to Madame de Tessé, 20 March 1787, in Jefferson, *Papers,* 11:226–28.

77. TJ to James Madison, 20 September 1785, ibid., 8:534–37.

78. TJ to Madame de Tessé, 20 March 1787, ibid., 11:227.

79. Ibid., 226.

80. For a journey in the other direction, from the past to the future, compare the reflective passage following the "memorable fable of the SEVEN SLEEPERS," in Gibbon, *Decline and Fall,* 3:363–85. TJ may have had this passage in mind, since his letter, like Gibbon's chapter, refers to the Vandals in particular, as will be discussed below. For further eighteenth-century instances of uchronia, see Louis-Sébastien Mercier's science-fiction novel *L'an 2440,* which TJ mentioned in a letter to Abigail Adams of 11 January 1817. *Adams-Jefferson Letters,* 504.

81. TJ to Madame de Tessé, 20 March 1787, in Jefferson, *Papers,* 11:226–28.

82. Madame de Tessé to TJ, 30 March 1787, ibid., 11:257–60.

83. TJ to Madame de Tott, 28 February 1787, ibid., 11:187.

84. For the persuasive argument of a prevalence of cyclical theories of history in the South during the revolutionary period, contrasting with New England millennialism, see Shalev, *Rome Reborn on Western Shores,* ch. 3. While TJ did not exemplify this trend, he was very well able to express himself in this mode, especially in the context of French and Roman history.

85. On Grégoire's invention of "un langage révolutionnaire de la conservation," see Vidler, "Grégoire, Lenoir," esp. 131–40. TJ later would employ the term *vandalism* when British troops burned the presidential mansion during the War of 1812. See TJ to Marquis de Lafayette, 14 February 1815, in *Thomas Jefferson: Writings,* 1363.

86. Vidler, "Grégoire, Lenoir," 141.

87. TJ referred to Native Americans variously as "indians," "aborigines," "(red) men of America," or by the proper names of their tribes. I use TJ's terminology when I describe his perspective. On his attitude toward the Indians, see Sheehan, *Seeds of Extinction;* Onuf, *Jefferson's Empire,* 18–52; and Wallace, *Jefferson and the Indians.*

88. TJ to Benjamin Smith Barton, 21 September 1809, in Jefferson, *Papers: Retirement Series,* 1:555–57.

89. Under query 11, TJ repeatedly emphasized the necessity of more "search." Jefferson, *Notes,* 93, 96, 101.

90. Ibid., 100.

91. Ibid., 101.

92. Ibid., 102.

93. TJ to Ezra Stiles, 1 September 1786, in Jefferson, *Papers,* 10:316–18. See Jefferson, *Notes,* 102, 282n12, for TJ's relativization of this position.

94. Jefferson, *Notes,* 208n10, 62, 63.

95. See Horsman, *Race and Manifest Destiny,* 16–24.

96. TJ inconsistently addressed Indians as "brothers" or "children," e.g., in "To Brother John Baptist de Coigne," June 1781; "To the Brothers of the Choctaw Nation,"

17 December 1803; and "To the Wolf and People of the Mandan Nation," 30 December 1806, in *Thomas Jefferson: Writings,* 551–66. For TJ's views on Indians as "formed in mind as well as in body, on the same module with the 'Homo sapiens Europæus,'" see Jefferson, *Notes,* 62–65.

97. Jefferson, *Notes,* 97. TJ relativized under query 14 his low opinion of Indian art, in comparison with the talents of blacks, which he thought still inferior. Cf. ibid., 140.

98. K. Lehmann, *American Humanist,* 25.

99. Jefferson, *Notes,* 96–97.

100. Ibid., 62. During TJ's vice-presidency, this passage of the *Notes* was criticized harshly in the press. TJ defended himself, not by answering directly, but by more research, which he published as an *Appendix to the Notes on Virginia Relative to the Murder of Logan's Family* in 1800, reprinted in ibid., 226–58.

101. TJ to Governor Henry of Maryland, 31 December 1797, reprinted in the 1800 *Appendix.* Ibid., 227.

102. Ibid., 63.

103. In 1774, the year of Logan's speech to Lord Dunmore, similar speeches were made in European theaters, as far away, in terms of space and direct influence, as in the premiere of Goethe's *Götz von Berlichingen* in Berlin. In the last act of the historical drama, the "Storm and Stress" hero Götz, a German noble savage from the Middle Ages, dies, like Logan, defeated and without heir but accompanied by several memorable speeches alluding, as does Logan's, to the likely oblivion of posterity and to his death and the disappearance of his historical period (features that apparently fulfilled crucial needs of a late Enlightenment audience). See Goethe, *Götz von Berlichingen,* 110–11, act 5, scene titled "Gärtchen am Turm."

104. See Onuf, *Jefferson's Empire,* 21.

105. Jefferson, *Summary View,* in Jefferson, *Papers,* 1:134–35.

106. TJ to Brother John Baptist de Coigne, ca. 1 June 1781, ibid., 6:60–64.

107. TJ to James Madison, 6 September 1789, ibid., 15:392. TJ reiterated these ideas in his later years, e.g., in a letter to Samuel Kercheval of 12 July 1816, in *Thomas Jefferson: Writings,* 1395–1403.

108. See Sloan, "Earth Belongs in Usufruct"; and Onuf, "Every Generation is an 'Independant Nation.'"

109. TJ to James Madison, 6 September 1789, in Jefferson, *Papers,* 15:392.

110. TJ to John Breckinridge, 29 January 1800, ibid., 31:344–45.

111. Using the mortality tables of the Count de Buffon, TJ constructed a political generation whose majority would be living "18 years 8. months, or say 19" years after achieving the "ripe age" of 21 years. TJ to James Madison, ibid., 15:393–94.

112. Ibid., 395.

113. Ibid., 395–96.

114. TJ's conception of gradual progress left some room for temporary regression, as would become especially evident, in the American context, during the Missouri Crisis (see chapter 6). For the stadialist character of TJ's conception of progress, see esp. TJ to William Ludlow, 6 September 1824, in *Thomas Jefferson: Writings,* 1496.

115. For the expression *long succession of generations,* see a passage that TJ would later criticize in his *Autobiography,* in which Burke worried about the "loss" of chivalry. While not in every respect oppositional to TJ's ideas, Burke's generational

thought could sound almost like a direct criticism of TJ's lack of connectives (making a similar point to one made by Madison in his reply to TJ's letter, although not as politely): "By this unprincipled facility of changing the state as often . . . as there are floating fancies or fashions, the whole chain and continuity of the commonwealth would be broken. No one generation could link with the other. Men would become little better than the flies of a summer. . . . Thus the commonwealth itself would, in a few generations, crumble away, be disconnected into the dust and powder of individuality, and at length dispersed to all the winds of heaven." Burke, *Reflections on the Revolution,* 65, 81–82.

116. "To the Brothers of the Choctaw Nation," 17 December 1803, in *Thomas Jefferson: Writings,* 558–60.

117. TJ to James Madison, 6 September 1789, in Jefferson, *Papers,* 15:392.

118. Ibid., 397.

119. See TJ to Richard Price, 8 January 1789, ibid., 14:420–24; or TJ to John Adams, 30 August 1787, in *Adams-Jefferson Letters,* 196.

120. According to Dumas Malone, TJ "commented rather more favorably on the course of the Revolution to Dr. Richard Price and Thomas Paine than he did to John Jay or even James Madison. The facts were the same but the nuances were slightly different. He expressed his hopes rather more freely on the one hand, and his fears on the other." Malone, *Jefferson and His Time,* 2:232.

121. For an example of TJ's shifting use of the term *revolution* in the same text, see TJ to John Jay, 9 May 1789, in Jefferson, *Papers,* 15:110–13.

122. See TJ to Comte Diodati, 3 August 1789, or TJ to Madame d'Enville, 2 April 1790, ibid., 15:325–27, 16:290–91.

123. TJ to James Madison, 28 August 1789, ibid., 15:364–69.

124. Not only did TJ make the (by then relatively commonplace) argument that the American "example" had inspired the French revolutionaries, he even thought that the success of the new French example would be "necessary to stay up our own, and to prevent it from falling back to that kind of Half-way house, the English constitution." See TJ to Richard Price, 8 January 1789, and TJ to George Mason, 4 February 1791, ibid., 14:420–24, 19:241–43.

125. For this argument, see his notorious "Adam and Eve" letter to William Short, 3 January 1793, ibid., 25:14–17.

126. For *prophecy* and *prognostications* in the context of the French Revolution, see TJ to Richard Price, 8 January 1789; TJ to Comte Diodati, 3 August 1789; and TJ to Madame d'Enville, 2 April 1790, ibid., 14:420–24, 15:325–27, 16:290–91.

127. TJ to George Washington, 4 December 1788, ibid., 14:328–32; Jefferson, "Autobiography," in *Thomas Jefferson: Writings,* 92.

128. TJ to John Jay, 9 May 1789, in Jefferson, *Papers,* 15:110–13; Jefferson, "Autobiography," in *Thomas Jefferson: Writings,* 95.

129. TJ to James Madison, 28 August 1789, in Jefferson, *Papers,* 15:364–69.

130. For the transitional concepts of nationhood in this period, see Onuf and Onuf, *Nations, Markets, and War,* esp. 144–72, 219–46.

131. See TJ to J. Hector St. John de Crèvecoeur, 9 August 1788, in Jefferson, *Papers,* 13:485–87; TJ to George Washington, 4 December 1788, ibid., 14:328–32; TJ to John Jay, 9 May 1789, ibid., 15:110–13; TJ to Comte Diodati, 3 August 1789, ibid., 15:325–27;

TJ to John Adams, 10 May 1789, *Adams-Jefferson Letters,* 237; and for a retrospective estimate of this impossibility, TJ to Benjamin Austin, 9 January 1816, in *Thomas Jefferson: Writings,* 1371.

132. On the complex phenomenon of historical acceleration in times of crisis, see Koselleck, *Zeitschichten,* 150–76, 177–202. For historical acceleration at the turn of the next century, see Adams, "A Law of Acceleration (1904)," in *Education of Henry Adams,* 407–14.

133. TJ to James Madison, 9 June 1793, in Jefferson, *Papers,* 26:239–42.

134. Thomas Paine, *The American Crisis,* no. 3, in Paine, *Collected Writings,* 116.

135. TJ to John Adams, 30 August 1787, *Adams-Jefferson Letters,* 196.

136. TJ to Maria Cosway, 25 July 1789, in Jefferson, *Papers,* 15:205.

137. TJ to Dugald Stewart, 26 April 1824, in *Thomas Jefferson: Writings,* 1487–89.

138. Jefferson, *Summary View,* in Jefferson, *Papers,* 1:125. In this text, TJ created an effect of acceleration also in the structure of his argument: the time of his writing is antiproportional to the time of the events it refers to, and he needs much more time to describe what took place in a few years than to describe what took place in more than a millennium.

139. "Future ages will hardly believe that the hardiness of one man adventured, *within the short compass of twelve years only,* to lay a foundation so broad & undisguised for tyranny over a people fostered & fixed in principles of freedom." In this phrase from the draft of the Declaration of Independence, the king is responsible for the accelerated pace of change, whereas the American people's "fixedness" promises future stability. TJ's draft quoted from Jefferson, "Autobiography," in *Thomas Jefferson: Writings,* 22, my emphasis.

140. See, e.g., TJ to John Adams, 28 December 1796 and 10 December 1819, in *Adams-Jefferson Letters,* 263, 548–49; and TJ to John Dickinson, 6 March 1801, TJ to Joseph Priestley, 21 March 1801, and TJ to Caesar A. Rodney, 10 February 1810, in *Thomas Jefferson: Writings,* 1084, 1086, 1217.

141. See Arendt, *On Revolution,* esp. 42–43.

142. On this dichotomy in TJ's thought, see the conclusion of query 18 in Jefferson, *Notes,* 163. For the expectation of significant peaceful change within his own lifetime, see TJ to Richard Price, 7 August 1785, in Jefferson, *Papers,* 8:357. See also TJ to Edward Bancroft, 16 January 1788, in which TJ untypically claimed to be making immediatist emancipation plans (although this extraordinary position may have been related to suspicions of Bancroft's status as British spy). Ibid., 14:492.

143. TJ to William A. Burwell, 28 January 1805, in Jefferson, *Works,* 10:126.

144. TJ to St. George Tucker, 28 August 1797, in Jefferson, *Papers,* 29:519–20.

145. On TJ's use of this topos, see H. Spahn, *Jefferson und die Sklaverei,* 39–42.

146. On St. George Tucker and the lack of success of his *Dissertation on Slavery* (1796) in Virginia, see McColley, *Slavery and Jeffersonian Virginia,* 134–35; and W. Jordan, *White over Black,* 555–60. For "apathy" even among Tucker's correspondents in Massachusetts, see Newman, *Transformation of American Abolitionism,* 36–37.

147. On TJ's relative silence on Haiti, see Rossignol, *Nationalist Ferment,* esp. 128–36.

148. Banning, *Jeffersonian Persuasion,* 208–70.

149. See the diary of Pennsylvania senator William Maclay, as quoted in Freeman, *Affairs of Honor,* 26.

150. For a concise interpretation of TJ's personal ambivalence toward party conflicts, see Bernstein, *Thomas Jefferson,* 81–104, 117–34.

151. For TJ's epistemological metaphors in this context, see, e.g., TJ to John Taylor, 4 June 1798, and TJ to Thomas Lomax, 12 March 1799, in Jefferson, *Papers,* 30:387–90, 31:244–45.

152. On the emerging "myth," in this period, of a qualitative contrast between the two revolutions, see Wood, *Radicalism of the American Revolution,* 231.

153. For this crucial connection to the "spirit of 1776," see Onuf, *Jefferson's Empire,* esp. 97–102.

154. TJ to John Breckinridge, 29 January 1800, in Jefferson, *Papers,* 31:344–45. Cf. TJ to John Taylor, 4 June 1798, or TJ to Thomas Lomax, 2 March 1799, ibid., 30:389, 31:244–45.

155. Jefferson, first inaugural address, 4 March 1801, in *Thomas Jefferson: Writings,* 492–96.

156. TJ to James Madison, 29 June 1792, in Jefferson, *Papers,* 24:133–34.

157. TJ to Benjamin Austin, 9 January 1816; Jefferson, "The Anas. 1791–1806"; and Jefferson, "Autobiography," in *Thomas Jefferson: Writings,* 1371, 673–74, 93.

158. Jefferson, "The Anas. 1791–1806," ibid., 674.

159. For the idea of a "republican millennium" from TJ's first inaugural address on, see Onuf, *Jefferson's Empire,* 107–8; and Onuf and Onuf, *Nations, Markets, and War,* esp. 221–25. More generally on American millennialism in this period, see Bloch, *Visionary Republic;* and Tuveson, *Redeemer Nation.*

160. Jefferson, first inaugural address, in *Thomas Jefferson: Writings,* 492.

161. See Onuf, "To declare them a free and independant people."

6. Beyond Example?

1. John Adams to TJ, 1 January 1812, and TJ to John Adams, 21 January 1812, in *Adams-Jefferson Letters,* 290, 290–92.

2. Cf. TJ to John Adams, 25 March 1826, ibid., 613–14.

3. Jefferson, first inaugural address, 4 March 1801, in *Thomas Jefferson: Writings,* 492.

4. TJ to John Adams, 21 January 1812, in *Adams-Jefferson Letters,* 291.

5. Ibid.

6. See, e.g., TJ to John Norvell, 14 June 1807, in *Thomas Jefferson: Writings,* 1177–78.

7. For a negative criticism of the timeless appeal of these writers, cf. Adams's reply to Jefferson's letter of 21 January 1812: "I have read Thucidides and Tacitus, so often, and at such distant Periods of my Life, that elegant, profound and enchanting as is their Style, I am weary of them. When I read them I seem to be only reading the History of my own Times and my own Life." John Adams to TJ, 3 February 1812, in *Adams-Jefferson Letters,* 295.

8. "I soon after [sitting for Gilbert Stuart's oil paintings] got him to sketch me in the medallion form, which he did on paper with crayons. Although a slight thing I gave him another hundred dollars, probably the treble of what he would have asked. This I

have; it is a very fine thing, although very perishable." TJ to Henry Dearborn, 5 July 1819, in *Writings of Thomas Jefferson,* 19:271. TJ apparently wished this precious but "perishable" image to be preserved, as he loaned it in 1815 to William Thornton, who copied it in different versions in crayon and oil. See Stein, *Worlds of Thomas Jefferson,* 141.

9. Ibid.

10. TJ to Nathaniel Burwell, 14 March 1818, in *Thomas Jefferson: Writings,* 1411–13; this letter is discussed below.

11. TJ to Judge Spencer Roane, 6 September 1819, ibid., 1425. For recent interpretations of the "Revolution of 1800," see Onuf, *Jefferson's Empire,* 80–108; and Horn, Lewis, and Onuf, *Revolution of 1800.*

12. Jefferson, first inaugural address, 4 March 1801, in *Thomas Jefferson: Writings,* 493.

13. TJ to Joseph Priestley, 21 March 1801, ibid., 1086.

14. See Eccles. 1:9–10. For the likely influence of cyclical concepts of time from Hellenistic culture on this relatively late part of the Old Testament, see Williams, "Proverbs and Ecclesiastes." By the time of his retirement, TJ felt absolutely certain that Montesquieu's concept of republican space did not apply to America. See, e.g., TJ to François Marbois, 14 June 1817, in *Thomas Jefferson: Writings,* 1409–11.

15. TJ to Joseph Priestley, 21 March 1801, in *Thomas Jefferson: Writings,* 1086.

16. See, e.g., TJ to John Dickinson, 6 March 1801: "But the band is removed, and they now see for themselves." Ibid., 14 June 1817, 1084–85.

17. As TJ frequently pointed out, he hoped Americans would refuse "to look backwards, not forwards, for improvement." TJ to Joseph Priestley, 21 March 1801, ibid., 1085. See also TJ to Elbridge Gerry, 26 January 1799, TJ to William Green Munford, 18 June 1799, and TJ to Joseph Priestley, 27 January 1800, in Jefferson, *Papers,* 30:645–50, 31:126–30, 339–41.

18. John Adams to TJ, 14 June 1813, in *Adams-Jefferson Letters,* 329–30.

19. TJ to Alexander von Humboldt, 6 December 1813, in *Thomas Jefferson: Writings,* 1311–14.

20. See TJ to Comte Diodati, 29 March 1807, ibid., 11:181–82, which establishes a similar contrast between European wars and American peace.

21. TJ to William Short, 28 November 1814, ibid., 1355–60. TJ could accept the evils of war within his conception of history as a natural catastrophe: "We must meet, therefore, the former as the casualties of tempests and earthquakes, and like them necessarily resulting from the constitution of the world" (1358).

22. On the postponement of the millennium in TJ's thought, see Onuf and Onuf, *Nations, Markets, and War,* 219–46.

23. TJ to Benjamin Austin, 9 January 1816, in *Thomas Jefferson: Writings,* 1369–72.

24. TJ to Roger C. Weightman, 24 June 1826, ibid., 1516–17.

25. TJ to John Adams, 10–11 August 1815, in *Adams-Jefferson Letters,* 453. On TJ's opinion of Marshall's *Life of Washington,* see Freeman, *Affairs of Honor,* 62–66; Burstein, *Jefferson's Secrets,* 211–22; and Cogliano, *Reputation and Legacy,* 49–61. Freeman's phrase "biographical war" appears adequate for describing TJ's perspective, as he considered it a characteristic of civilized societies that the "pen" replaced

the "sword." See, e.g., TJ to Thomas Paine, 19 June 1792, in Jefferson, *Papers,* 20:312–13.

26. For a comparison of their memoirs, see Sloan, "Presidents as Historians," 266–77.

27. For TJ's disappointment, see TJ to Joel Barlow, 16 April 1811, in *Writings of Thomas Jefferson,* 13:44–45. After his return from Europe in 1804, Barlow instead produced another version of his 1787 poem *The Vision of Columbus,* namely, *The Columbiad: A Poem,* published in 1807. For Barlow's preference of exemplary poetry over exemplary historiography, see Lang, "Joel Barlows 'Dissertation.' "

28. See, e.g., TJ to Justice William Johnson, 12 June 1823, in *Thomas Jefferson: Writings,* 1469–77.

29. See, e.g., TJ to Dr. Josephus B. Stuart, 10 May 1817, and TJ to Horatio Gates Spafford, 11 May 1819, in *Writings of Thomas Jefferson,* 15:111–14, 189–91.

30. Among these, Madison was foremost. See TJ to James Madison, 17 February 1826, in *Thomas Jefferson: Writings,* 1512–15.

31. See, e.g., TJ to L. H. Girardin, 27 March 1815: "As to what is to be said of myself, I of course am not the judge. But my sincere wish is that the faithful historian, like the able surgeon, would consider me in his hands, while living, as a dead subject, that the same judgment may now be expressed which will be rendered hereafter, so far as my small agency in human affairs may attract future notice; and I would of choice now stand as at the bar of posterity." *Writings of Thomas Jefferson,* 14:295. Cf. TJ to John Adams, 27 June 1813, in *Adams-Jefferson Letters,* 337.

32. TJ to John Adams, 15 June 1813. This letter prompted a similar disclaimer in John Adams to TJ, 25 June 1813: "You must have misunderstood me. . . . Your Letters will do you no dishonour. As to mine I care not a farthing. My Reputation has been so much the Sport of the public for fifty Years, and will be with Posterity, that I hold it, a bubble, a Gossameur, that idles in the wanton Summers Air." *Adams-Jefferson Letters,* 332–33, 333.

33. *Thomas Jefferson: Writings,* 3, 663.

34. Cogliano, *Reputation and Legacy,* 55.

35. Jefferson, "Autobiography," in *Thomas Jefferson: Writings,* 97.

36. See Wood, "Conspiracy and the Paranoid Style," esp. 306–8.

37. Jefferson, "Autobiography," in *Thomas Jefferson: Writings,* 37.

38. For an exception, see TJ's account of a dinner in Paris in 1789 to which he invited leading French Patriots. At that dinner they achieved a compromise on the French constitution. Ibid., 95–96.

39. A small degree of self-historicization may be found in TJ's description of the colonial legislature when he became a member in 1769. Ibid., 5.

40. Ricoeur, *Zeit und Erzählung,* 3:96–157. For TJ's use of the term *chronicle,* see his letter to William Duane of 12 August 1810, in *Thomas Jefferson: Writings,* 1229.

41. He often cited his own letters as sources, for example: "[For a sketch of the life & character of Mr. Wythe see my letter of Aug. 31. 20 to Mr. John Saunderson.]" Jefferson, "Autobiography," ibid., 5.

42. Jefferson, explanations to the "Anas," ibid., 662.

43. For TJ's nineteenth-century use of the storm metaphor, see TJ to Caesar A.

Rodney, 10 February 1810, ibid., 1217; and TJ to John Adams, 10 December 1819, in *Adams-Jefferson Letters,* 548–49.

44. TJ to William Duane, 12 August 1810, in *Thomas Jefferson: Writings,* 1228. For similar metaphors, see TJ to Doctor Walter Jones, 5 March 1810, in Jefferson, *Works,* 11:137–39.

45. Locke, *Essay Concerning Human Understanding,* 108 (2.14.12).

46. Ibid., 105 (2.14.4).

47. TJ also used this metaphor to describe the "sleep of despotism" under the *ancien régime.* This use was consistent with TJ's positive use of the metaphor for his retirement, referring in each case to the absence of active republican politics. For the "sleep of despotism," see the *Autobiography* or TJ to Philip Mazzei, 24 April 1796, in Jefferson, *Papers,* 29:81–83, whose unauthorized publication caused a scandal during TJ's vice-presidency.

48. TJ to Benjamin Austin, 9 January 1816, in *Thomas Jefferson: Writings,* 1370.

49. TJ to William Short, 5 May 1816, in Series 1, General Correspondence, 1651–1827, *Jefferson Papers, 1606–1827,* transcribed by Gerard W. Gawalt, Manuscript Division, Library of Congress. On TJ's accepting attitude toward his approaching death, see Burstein, *Jefferson's Secrets,* esp. 1–15, 274–88.

50. For one of many examples, see TJ to Horatio Gates Spafford, 11 May 1819: "In no circumstance has nature been kinder to us, than in the soft gradations by which she prepares us to part willingly with what we are not destined to retain. First one faculty is withdrawn, and then another, sight, hearing, memory, accuracy, affections, & friends, filched one by one, till we are left among strangers." *Writings of Thomas Jefferson,* 15:189.

51. TJ was especially averse to investing energy in assisting foreigners in their historical studies of the American republic, perhaps as a result of his disappointing experiences with Soulés and Mayer in the 1780s. He usually referred them to other Americans. See, e.g., TJ to Count John Potocki, 12 May 1811: "Time tells me I am nearly done with the history of the world; that I am now far advanced in the last chapter of my own, and that its last verse will be read out ere a few letters could pass between St. Petersburg and Monticello. I shall serve you therefore more permanently, by bequeathing to you another correspondent, more able, more industrious, and more likely to continue in life than myself." Ibid., 13:61–62.

52. See, e.g., TJ to Thomas Jefferson Smith, 21 February 1825: "This letter will, to you, be as one from the dead. The writer will be in the grave before you can weight its counsels." *Thomas Jefferson: Writings,* 1499–1500.

53. TJ to Nathaniel Macon, 12 January 1819, in *Writings of Thomas Jefferson,* 15:179–81.

54. TJ to John Armstrong, 6 March 1809, in Jefferson, *Papers: Retirement Series,* 1:20.

55. TJ to Horatio Gates Spafford, 11 May 1819, in *Writings of Thomas Jefferson,* 15:189.

56. See, e.g., TJ to John Adams, 1 August 1816: "When we have lived our generation out, we should not wish to encroach on another." *Adams-Jefferson Letters,* 484.

57. See TJ to Samuel Kercheval, 12 July 1816: "It is now forty years since the constitution of Virginia was made. The same tables inform us, that, within that period,

two-thirds of the adults then living are now dead. Have then the remaining third, even if they had the wish, the right to hold in obedience to their will, and to laws heretofore made by them, the other two-thirds . . . ?" *Thomas Jefferson: Writings,* 1402.

58. In his letter to Samuel Kercheval, TJ described "our inexperience of self-government" in the late 1770s, which "occasioned gross departures . . . from genuine republican canons. In truth, the abuses of monarchy had so much filled all the space of political contemplation, that we imagined everything republican which was not monarchy." Ibid., 1396. See also TJ to Benjamin Austin, 9 January 1816: "You tell me I am quoted by those who wish to continue our dependence on England for manufactures. There was a time when I might have been so quoted with more candor, but within the thirty years which have since elapsed, how are circumstances changed!" Ibid., 1370.

59. For TJ's reaction to the Missouri Crisis, see Onuf, *Jefferson's Empire,* 109–46; Malone, *Jefferson and His Time,* 6:328–61; and J. Miller, *Wolf by the Ears,* 221–52.

60. TJ to John Holmes, 22 April 1820, in *Thomas Jefferson: Writings,* 1433–35.

61. TJ to John Adams, 22 January 1821, in *Adams-Jefferson Letters,* 569–70.

62. TJ to John Holmes, 22 April 1820, in *Thomas Jefferson: Writings,* 1434–35.

63. TJ to John Adams, 5 July 1814, in *Adams-Jefferson Letters,* 434.

64. TJ to Edward Coles, 25 August 1814, in *Thomas Jefferson: Writings,* 1343–46; TJ to William Green Munford, 18 June 1799, in Jefferson, *Papers,* 31:126–30.

65. In Hesiod's description of five generations, from a golden to an iron one, the fourth is the only generation that progresses compared with its predecessor, as it is "nobler and more righteous, a god-like race of hero-men who are called demi-gods," followed by the regression of the present (iron) age. TJ, of course, did not share the sense of general decline, but his conception of the relationship between his father's, his own, and his children's generations is remarkably similar to the relationship between Hesiod's last three generations: TJ spoke of his own generation both as "demigods" and as a "Heroic age." TJ to John Adams, 30 August 1787 and 25 March 1826, in *Adams-Jefferson Letters,* 196, 614. Cf. Hesiod, *Works and Days,* pp. 12–15 (lines 156–69).

66. Earlier in the letter to Holmes, TJ had used the image of the balance for the problem of slavery: "But as it is, we have the wolf by the ears, and we can neither hold him, nor safely let him go. Justice is in one scale, and self-preservation in the other." TJ to John Holmes, 22 April 1820, in *Thomas Jefferson: Writings,* 1434.

67. On the musical collection of the Jefferson family, which incidentally contained sheets of *Don Giovanni,* see Cripe, *Thomas Jefferson and Music,* 105–6.

68. TJ to John Adams, 22 January 1821, in *Adams-Jefferson Letters,* 569–70.

69. TJ to Marquis de Lafayette, 14 February 1815, in *Thomas Jefferson: Writings,* 1360–66.

70. TJ to John Adams, 22 January 1821, in *Adams-Jefferson Letters,* 569. For a compelling interpretation of these metaphors from European politics, see Onuf, *Jefferson's Empire,* 111, 121–22.

71. Irving, *History of New York,* 166.

72. TJ to John Adams, 17 May 1818, in *Adams-Jefferson Letters,* 524; TJ to President Monroe, 27 October 1823, in ; *Thomas Jefferson: Writings,* 1481.

73. Shelley, "Time" (written 1821, published 1824).

74. TJ to Abigail Adams, 11 January 1817: "But those 20 years, alas! where are they?

With those beyond the flood. Our next meeting must then be in the country to which they have flown." *Adams-Jefferson Letters,* 504.

75. TJ to Albert Gallatin, 26 December 1820, in *Thomas Jefferson: Writings,* 1447–50.

76. TJ to John Adams, 10 December 1819, in *Adams-Jefferson Letters,* 548–50.

77. Ibid.

78. See, e.g., TJ to John Adams, 5 July 1814, ibid., 433; and TJ to Francis Wayles Eppes, 19 January 1821, in Jefferson, *Family Letters,* 438.

79. TJ to John Adams, 5 July 1814, in *Adams-Jefferson Letters,* 430–34.

80. See Walton, "Hume and Jefferson."

81. TJ to William Duane, 12 August 1810, in *Thomas Jefferson: Writings,* 1229.

82. TJ to John Norvell, 14 June 1807, ibid., 1176–79.

83. Ibid.

84. For different interpretations see Malone, *Jefferson and His Time,* 6:204–7; and Levy, *Jefferson and Civil Liberties,* 144–46.

85. TJ to Major John Cartwright, 4 June 1824, in *Thomas Jefferson: Writings,* 1491.

86. TJ to William Duane, 12 August 1810: "I cannot say that his amendments are either in matter or manner in the fine style of Hume. . . . Unfortunately he has abridged Hume, by leaving out all the less important details." Ibid., 1227–31.

87. TJ to George Washington Lewis, 25 October 1825, in *Writings of Thomas Jefferson,* 16:125–26.

88. TJ to William Duane, 12 August 1810, in *Thomas Jefferson: Writings,* 1227–31.

89. TJ to George Washington Lewis, 25 October 1825, in *Writings of Thomas Jefferson,* 16:125–26.

90. TJ to F. A. van der Kemp, 22 March 1812, ibid., 13:135–37.

91. The aged Jefferson, as J. Jefferson Looney has recently pointed out, probably continued to read novels for pleasure himself but, in contrast to in his early years, failed to acknowledge this reading consistently in his book lists and in the catalog of the library he sold to Congress. I thank Mr. Looney for letting me read his conference paper "I shall not retain a single one."

92. See, e.g., TJ to Justice William Johnson, 27 October 1822, in *Thomas Jefferson: Writings,* 1459–63.

93. TJ to Robert Skipwith, 3 August 1771, in Jefferson, *Papers,* 1:76–78.

94. TJ to Nathaniel Burwell, 4 March 1818, in *Thomas Jefferson: Writings,* 1411–13. For a discussion of TJ's thoughts on young women's reading, see Burstein, *Jefferson's Secrets,* 87–110.

95. Jefferson, "To the Inhabitants of Albermarle County, in Virginia," 3 April 1809, in *Thomas Jefferson: Writings,* 550.

96. Malone, *Jefferson and His Time,* 6:365–425; Hellenbrand, *Unfinished Revolution,* esp. 150–53; Wills, *Mr. Jefferson's University.*

97. W. Adams, *Jefferson's Monticello;* McLaughlin, *Jefferson and Monticello;* Stein, *Worlds of Thomas Jefferson;* Isaac, "First Monticello."

98. Peterson, *Jefferson Image,* 380–94; Cogliano, *Reputation and Legacy,* 106–36.

99. On this irony, see Peterson, *Jefferson Image,* 388–94.

100. For this definition of the aesthetic object, see Fluck, "Pragmatism and Aesthetic Experience."

101. TJ to Horatio Gates Spafford, 11 May 1819, in *Writings of Thomas Jefferson,* 15:189.

102. On the recent development of the nickel, see Cogliano, *Reputation and Legacy,* 106–7.

103. William Wirt, eulogy, 19 October 1826, in *Writings of Thomas Jefferson,* 13:xxiii.

Epilogue

1. TJ to Roger C. Weightman, 24 June 1826, in *Thomas Jefferson: Writings,* 1516–17, quotation on 1517.

2. TJ to Doctor Thomas Humphreys, 8 February 1817: "I leave it, therefore, to time, and not at all without hope that the day will come, equally desirable and welcome to us as to them." Jefferson, *Works,* 12:53–55.

3. TJ to John Adams, 1 August 1816, in *Adams-Jefferson Letters,* 485.

4. TJ to David Barrow, 1 May 1815, in Jefferson, *Works,* 11:470–71.

5. Jefferson, *Notes,* 143.

6. TJ to Edward Coles, 25 August 1814, in *Thomas Jefferson: Writings,* 1343–46.

7. TJ to John Adams, 25 March 1826, in *Adams-Jefferson Letters,* 614.

8. TJ to William Green Munford, 18 June 1799, in Jefferson, *Papers,* 31:126–30.

9. TJ to St. George Tucker, 28 August 1797, ibid., 29:519–20.

10. Tocqueville, *Democracy in America,* 831.

11. TJ to William Ludlow, 6 September 1824, in *Thomas Jefferson: Writings,* 1497.

12. TJ to John Melish, 10 December 1814, in *Writings of Thomas Jefferson,* 14:220.

Bibliography

Adair, Douglass. *Fame and the Founding Fathers: Essays by Douglass Adair.* Ed. Trevor Colbourn. New York: Norton, 1974.

Adams, Henry. *The Education of Henry Adams.* Ed. Ira B. Nadel. Oxford: Oxford UP, 1999.

Adams, John. "Autobiography." In *Diary and Autobiography of John Adams,* ed. Lyman H. Butterfield, 3:253–449. Cambridge, MA: Harvard UP, 1963.

———. "Defence of the Constitutions of the United States of America." In *The Works of John Adams,* ed. Charles Francis Adams, 4:271–588. Boston: Little, Brown, 1856.

Adams, William H., ed. *The Eye of Thomas Jefferson.* Washington, DC: National Gallery of Art, 1976. Exhibition catalog.

———. *Jefferson's Monticello.* With Principal Photography by Langdon Clay. New York: Abbeville, 1983.

———. *The Paris Years of Thomas Jefferson.* New Haven, CT: Yale UP, 1997.

Allen, Thomas M. *A Republic in Time: Temporality and Social Imagination in Nineteenth-Century America.* Chapel Hill: U of North Carolina P, 2008.

Arendt, Hannah. *On Revolution.* London: Faber & Faber, 1963.

Aristotle. *The Poetics.* Trans. Hamilton Fyfe. Loeb Classical Library. 1927. Cambridge, MA: Harvard UP; London: Heinemann, 1973.

———. *The Politics.* Trans. Harris Rackham. Loeb Classical Library. Cambridge, MA: Harvard UP; London: Heinemann, 1932.

Augustinus, Aurelius. *Confessiones/Confessions.* Trans. Henry Chadwick. Oxford: Oxford UP, 1991.

Austen, Jane. *Mansfield Park.* London: Penguin Classics, 1996.

Bacon, Francis. *The Works of Francis Bacon.* Ed. James Spedding et al. 14 vols. London: Longmans, 1857–74. Reprint, Stuttgart-Bad Cannstadt: Frommann, 1989.

Banning, Lance. *The Jeffersonian Persuasion: Evolution of a Party Ideology.* Ithaca, NY: Cornell UP, 1978.

Baudelaire, Charles. *Les fleurs du mal.* Ed. Jacques Dupont. Paris: Flammarion, 1991.

Bear, James A., Jr., ed. *Jefferson at Monticello.* Charlottesville: UP of Virginia, 1967.

Bedini, Silvio A. "Thomas Jefferson, Clock-Designer." *Proceedings of the American Philosophical Society* 108, no. 3 (1964): 163–80.

——. *Thomas Jefferson: Statesman of Science.* New York: Macmillan, 1990.

Berlin, Isaiah. *The Proper Study of Mankind: An Anthology of Essays.* Ed. Henry Hardy and Roger Hausheer. London: Random House, 1997.

Bernstein, Richard B. *Thomas Jefferson.* Oxford: Oxford UP, 2003.

Bloch, Ruth H. *Visionary Republic: Millennial Themes in American Thought, 1756–1800.* Cambridge: Cambridge UP, 1985.

Blumenberg, Hans. *Lebenszeit und Weltzeit.* Frankfurt: Suhrkamp, 1986.

Bolingbroke, Henry St. John. *Letters on the Study and Use of History, Etc.* London: Ward, Lock, 1880.

Braudy, Leo. *Narrative Form in History and Fiction: Hume, Fielding, and Gibbon.* Princeton, NJ: Princeton UP, 1970.

Breen, T. H. *Puritans and Adventurers: Change and Persistence in Early America.* New York: Oxford UP, 1980.

Brown, Charles Brockden. *Wieland, or The Transformation.* Amherst, NY: Prometheus Books, 1997.

Brunner, Otto, Werner Conze, and Reinhart Koselleck, eds. *Geschichtliche Grundbegriffe: Historisches Lexikon zur politisch-sozialen Sprache in Deutschland.* 7 vols. Stuttgart: Klett, 1972–97.

Burke, Edmund. *Reflections on the Revolution in France.* New Haven, CT: Yale UP, 2003.

Burnett, Edmund C., ed. *Letters of Members of the Continental Congress.* 8 vols. Gloucester, MA: P. Smith, 1963.

Burstein, Andrew. *The Inner Jefferson: Portrait of Grieving Optimist.* Charlottesville: UP of Virginia, 1995.

——. *Jefferson's Secrets: Death and Desire at Monticello.* New York: Basic Books, 2005.

Burstein, Andrew, and Catherine Mowbray. "Jefferson and Sterne." *Early American Literature* 29 (1994): 19–34.

Cicero. *De Oratore in Two Volumes.* Trans. E. W. Sutton. Vols. 3–4 of *Cicero in Twenty-Eight Volumes.* Loeb Classical Library. Cambridge, MA: Harvard UP; London: Heinemann, 1976.

Cogliano, Francis D. *Thomas Jefferson: Reputation and Legacy.* Edinburgh: Edinburgh UP, 2006.

Cohen, Lester H. *Revolutionary Histories: Contemporary Narratives of the American Revolution.* Ithaca, NY: Cornell UP, 1980.

Colbourn, H. Trevor. *The Lamp of Experience: Whig History and the Intellectual Origins of the American Revolution.* 1965. Indianapolis: Liberty Fund, 1998.

——. "Thomas Jefferson's Use of the Past." *William and Mary Quarterly* 15 (1958): 56–70.

Collingwood, R. G. *The Idea of History.* 1946. Oxford: Oxford UP, 1968.

Condorcet, Marquis de [M. J. A. Nicolas de Caritat]. *Œuvres.* Ed. Arthur O'Connor and Dominique Arago. 12 vols. Stuttgart-Bad Cannstadt: Frommann, 1968.

Conrad, Stephen A. "Putting Rights Talk in Its Place: The *Summary View* Revisited." In Onuf, *Jeffersonian Legacies,* 254-80.

Conze, Werner, and Antje Sommer. "Rasse." In Brunner, Conze, and Koselleck, *Geschichtliche Grundbegriffe,* 5:135-78.

Cripe, Helen. *Thomas Jefferson and Music.* Charlottesville: UP of Virginia, 1974.

Curti, Merle. *Human Nature in American Thought: A History.* Madison: U of Wisconsin P, 1980.

Davis, David Brion. "New Sidelights on Early Antislavery Radicalism." *William and Mary Quarterly* 28 (1971): 585-94.

——. *The Problem of Slavery in the Age of Revolution, 1770–1823.* Ithaca, NY: Cornell UP, 1975.

Defoe, Daniel. *The Life and Strange Surprizing Adventures of Robinson Crusoe of York, Mariner.* Ed. J. Donald Crowley. Oxford: Oxford UP, 1983.

Diderot, Denis. *Jacques le fataliste.* Ed. Pierre Chartier. Paris: Livre de Poche, 2000.

Dohrn-van Rossum, Gerhard. *History of the Hour: Clocks and Modern Temporal Orders.* Trans. Thomas Dunlap. Chicago: U of Chicago P, 1996.

——. "Zeit der Kirche–Zeit der Händler–Zeit der Städte: Die mechanische Uhr und der Wandel des Zeitbewußtseins im Spätmittelalter." In *Zerstörung und Wiederaneignung von Zeit,* ed. Rainer Zoll, 89-119. Frankfurt: Suhrkamp, 1988.

Douglass, Frederick. *Narrative of the Life of Frederick Douglass, an American Slave, Written by Himself.* Ed. William L. Andrews and William S. McFeely. New York: Norton, 1997.

Elias, Norbert. *Über den Prozeß der Zivilisation: Soziogenetische und psychogenetische Untersuchungen.* Vol. 2. Frankfurt: Suhrkamp, 1976.

Ellis, Joseph J. *American Sphinx: The Character of Thomas Jefferson.* New York: Knopf, 1996.

Emerson, Ralph Waldo. "Nature." In *The Complete Essays and Other Writings,* ed. Brooks Atkinson, 3-42. 1940. New York: Random House, 1950.

——. "Self-Reliance." In *The Complete Essays and Other Writings,* ed. Brooks Atkinson, 145-69. 1940. New York: Random House, 1950.

Engler, Bernd, and Kurt Müller. "Einleitung: Das Exempel und seine Funktionalisierungen." In Engler and Müller, *Exempla,* 9-20.

——. *Exempla: Studien zur Bedeutung und Funktion exemplarischen Erzählens.* Ed. Bernd Engler and Kurt Müller. Berlin: Duncker und Humblot, 1995.

Equiano, Olaudah. *The Interesting Narrative and Other Writings.* Ed. Vincent Carretta. Rev. ed. London: Penguin, 2003.

Fabian, Johannes. *Time and the Other: How Anthropology Makes Its Object.* New York: Columbia UP, 1983.

Faulkner, William. *Absalom, Absalom!* New York: Vintage, 1990.

Ferling, John. *Setting the World Ablaze: Washington, Adams, Jefferson, and the American Revolution.* Oxford: Oxford UP, 2000.

Fetscher, Iring. *Hegels Lehre vom Menschen: Kommentar zu den §§ 387 bis 482 der Philosophischen Wissenschaften.* Stuttgart-Bad Cannstadt: Frommann, 1970.

Fliegelman, Jay. *Declaring Independence: Jefferson, Natural Language, and the Culture of Performance.* Stanford, CA: Stanford UP, 1993.

——. *Prodigals and Pilgrims: The American Revolution against Patriarchal Authority, 1750–1800.* Cambridge: Cambridge UP, 1982.

Fluck, Winfried. *Das kulturelle Imaginäre: Eine Funktionsgeschichte des amerikanischen Romans, 1790–1900.* Frankfurt: Suhrkamp, 1997.

——. "Pragmatism and Aesthetic Experience." In *Pragmatism and Literary Studies,* ed. Winfried Fluck, 227–42. Tübingen: Narr, 1999.

Foster, E. A., et al. "Jefferson Fathered Slave's Last Child." *Nature,* 5 November 1998, 27–28.

Franklin, Benjamin. *The Autobiography and Other Writings by Benjamin Franklin.* Ed. Peter Shaw. New York: Bantam Books, 1982.

Freeman, Joanne B. *Affairs of Honor: National Politics in the New Republic.* New Haven, CT: Yale UP, 2001.

French, Scot A., and Edward L. Ayers. "The Strange Career of Thomas Jefferson: Race and Slavery in American Memory, 1943-1993." In Onuf, *Jeffersonian Legacies,* 418-56.

Genovese, Eugene D. *The Slaveholders' Dilemma: Freedom and Progress in Southern Conservative Thought.* Columbia: U of South Carolina P, 1992.

Gibbon, Edward. *The History of the Decline and Fall of the Roman Empire.* Ed. H. H. Milman. 6 vols. 1846. Reprint, Philadelphia: Lippincott, 1877.

Goethe, Johann Wolfgang. "Den Vereinigten Staaten." In Goethe, *Werke,* ed. Hendrik Birus and Karl Eibl, 1:255–57. Darmstadt: Wissenschaftliche Buchgesellschaft, 1998.

——. *Götz von Berlichingen mit der eisernen Hand.* Stuttgart: Reclam, 1993.

Gordon-Reed, Annette. *The Hemingses of Monticello: An American Family.* New York: Norton, 2008.

——. *Thomas Jefferson and Sally Hemings: An American Controversy.* Charlottesville: UP of Virginia, 1997.

Greene, Jack P. "The Intellectual Reconstruction of Virginia in the Age of Jefferson." In Onuf, *Jeffersonian Legacies,* 225–53.

Günther, Horst, and Reinhart Koselleck. "Geschichte, Historie." In Brunner, Conze, and Koselleck, *Geschichtliche Grundbegriffe,* 2:593–717.

Gurevich, A. J. "Time as a Problem in Cultural History." In *Cultures and Time,* ed. Louis Gardet et al., 229–45. Paris: UNESCO Press, 1976.

Guthrie, James R. *Above Time: Emerson's and Thoreau's Temporal Revolutions.* Columbia: U of Missouri P, 2001.

Hegel, Georg Wilhelm Friedrich. *Phänomenologie des Geistes.* Frankfurt: Ullstein, 1970.

Hellenbrand, Harold. *The Unfinished Revolution: Education and Politics in the Thought of Thomas Jefferson.* Newark: U of Delaware P, 1990.

Helo, Ari, and Peter S. Onuf. "Jefferson, Morality, and the Problem of Slavery." *William and Mary Quarterly* 60 (2003): 583–614.

Hentz, Caroline Lee. *The Planter's Northern Bride.* 1854. Reprint, Chapel Hill: U of North Carolina P, 1970.

Hesiod. *Theogony.* In *Hesiod: The Homeric Hymns and Homerica,* trans. Hugh G. Evelyn-White, 78–153. Loeb Classical Library. 1914. Cambridge, MA: Harvard UP; London: Heinemann, 1959.

——. *Works and Days.* In *Hesiod: The Homeric Hymns and Homerica,* trans. Hugh G. Evelyn-White, 2–65. Loeb Classical Library. 1914. Cambridge, MA: Harvard UP; London: Heinemann, 1959.

Hirschman, Albert O. *The Passions and the Interests: Political Arguments for Capitalism before Its Triumph.* Princeton, NJ: Princeton UP, 1977.

The Holy Bible: An Exact Reprint in Roman Type, Page for Page, of the Authorized Version Published in the Year 1611. 1833. Oxford: Oxford UP, 1985.

Home, Henry, Lord Kames. *Elements of Criticism.* 1840. Reprint, Honolulu: UP of the Pacific, 2002.

——. *Historical Law-Tracts.* Birmingham, AL: Legal Classics Library, 1988.

Horn, James, Jan Lewis, and Peter S. Onuf, eds. *The Revolution of 1800: Democracy, Race, and the New Republic.* Charlottesville: U of Virginia P, 2002.

Horsman, Reginald. *Race and Manifest Destiny: The Origins of American Racial Anglo-Saxonism.* Cambridge, MA: Harvard UP, 1981.

Hume, David. *An Enquiry concerning Human Understanding.* Ed. Tom L. Beauchamp. Oxford: Oxford UP, 1999.

——. *The History of England from the Invasion of Julius Caesar to the Revolution in 1688.* 6 vols. 1778. Reprint, Indianapolis: Liberty Classics, 1983–85.

——. *A Treatise of Human Nature.* 1888. Ed. L. A. Selby-Bigge. 2nd ed., with text revised and variant readings by P. H. Nidditch. Oxford: Clarendon, 1978.

Irving, Washington. *A History of New York, by Diedrich Knickerbocker.* Ed. Edwin T. Bowden. New Haven, CT: College & University Press; New York: Twayne, 1964.

Isaac, Rhys. "The First Monticello." In Onuf, *Jeffersonian Legacies,* 77–108.

Jayne, Allen. *Jefferson's Declaration of Independence: Origins, Philosophy, and Theology.* Lexington: U of Kentucky P, 1998.

Jefferson, Thomas. *The Commonplace Book of Thomas Jefferson: A Repertory of His Ideas on Government.* Ed. Gilbert Chinard. Baltimore: Johns Hopkins P, 1926.

——. *The Family Letters of Thomas Jefferson.* Ed. Edwin M. Betts and James A. Bear, Jr. 1966. Charlottesville: UP of Virginia, 1986.

——. *The Garden and Farm Books of Thomas Jefferson.* Ed. Robert C. Baron. Golden, CO: Fulcrum, 1987.

——. *Jefferson's Extracts from the Gospels: "The Philosophy of Jesus" and "The Life and Morals of Jesus."* Ed. Dickinson W. Adams and Ruth W. Lester. Princeton, NJ: Princeton UP, 1983.

——. *Jefferson's Literary Commonplace Book.* Ed. Douglas L. Wilson. Princeton, NJ: Princeton UP, 1989.

——. *Jefferson's Memorandum Books: Accounts, with Legal Records and Miscellany, 1767–1826.* Ed. James A. Bear, Jr., and Lucia C. Stanton. 2 vols. Princeton, NJ: Princeton UP, 1997.

——. *Notes on the State of Virginia.* Ed. William Peden. 1954. Chapel Hill: U of North Carolina P, 1982.

——. *The Papers of Thomas Jefferson.* Ed. Julian P. Boyd, Charles Cullen, John Catanzariti, and Barbara Oberg. 35 vols. to date. Princeton, NJ: Princeton UP, 1950–.

——. *The Papers of Thomas Jefferson: Retirement Series.* Ed. J. Jefferson Looney. 5 vols. to date. Princeton, NJ: Princeton UP, 2004–.

——. *Thomas Jefferson Papers: An Electronic Archive.* Boston: Massachusetts Historical Society, 2003. http://www.thomasjeffersonpapers.org/.

——. *The Thomas Jefferson Papers, 1606–1827.* Washington, DC: Library of Congress. http://www.memory.loc.gov/ammem/collections/jefferson_papers/.

——. *Thomas Jefferson's Architectural Drawings.* Ed. Frederick D. Nichols. 1960. Charlottesville: Thomas Jefferson Memorial Foundation / UP of Virginia, 2001.

——. *Thomas Jefferson: Writings.* Ed. Merrill D. Peterson. New York: Library of America, 1984.

——. *The Works of Thomas Jefferson.* Ed. Paul L. Ford. 1892–99. Federal ed. 12 vols. New York: Putnam's, 1904–5.

——. *The Writings of Thomas Jefferson.* Ed. Albert E. Bergh and Andrew A. Lipscomb. 20 vols. Washington, DC: Thomas Jefferson Memorial Association, 1903–5.

Jefferson, Thomas, Abigail Adams, and John Adams. *The Adams-Jefferson Letters: The Complete Correspondence between Thomas Jefferson and Abigail and John Adams.* Ed. Lester J. Cappon. 1959. Chapel Hill: U of North Carolina P, 1987.

Jennings, Edward M. "The Consequences of Prediction." *Studies on Voltaire & the Eighteenth Century* 153 (1976): 1131–50.

Jordan, Daniel P., et al. *Report of the Research Committee on Thomas Jefferson and Sally Hemings.* Charlottesville, VA: Thomas Jefferson Memorial Foundation, 2000. http://www.monticello.org/site/plantation-and-slavery/report-research-committee-thomas-jefferson-and-sally-hemings.

Jordan, Winthrop D. *White over Black: American Attitudes toward the Negro, 1550–1812.* Chapel Hill: U of North Carolina P, 1968.

Kant, Immanuel. *Kritik der reinen Vernunft.* Hamburg: Felix Meiner, 1956.

——. "Mutmasslicher Anfang der Menschengeschichte." In *Werke in zwölf Bänden,* ed. Wilhelm Weischedel, vol. 11, *Schriften zur Anthropologie, Geschichtsphilosophie, Politik und Pädagogik 1,* 85–102. Frankfurt: Suhrkamp, 1964.

Kelleter, Frank. *Amerikanische Aufklärung: Sprachen der Rationalität im Zeitalter der Revolution.* Paderborn: Schöningh, 2002.

Kennedy, Roger G. *Mr. Jefferson's Lost Cause: Land, Farmers, Slavery, and the Louisiana Purchase.* New York: Oxford UP, 2003.

Kerber, Linda. *Women of the Republic: Intellect and Ideology in Revolutionary America.* Chapel Hill: U of North Carolina P, 1980.

Kessel, Martina. *Langeweile: Zum Umgang mit Zeit und Gefühlen vom späten 18. bis zum frühen 20. Jahrhundert.* Göttingen: Wallstein, 2001.

Kindermann, Wolf. *Man Unknown to Himself: Kritische Reflexion der amerikanischen Aufklärung; Crèvecoeur, Benjamin Rush, Charles Brockden Brown.* Tübingen: Narr, 1993.

Koch, Adrienne. *The Philosophy of Thomas Jefferson.* New York: Columbia UP, 1943.

Kojève, Alexandre. *Hegel: Eine Vergegenwärtigung seines Denkens; Kommentar zur "Phänomenologie des Geistes."* Ed. Iring Fetscher. Trans. I. Fetscher et al. Frankfurt: Suhrkamp, 1975.

Koselleck, Reinhart. "Einleitung." In Brunner, Conze, and Koselleck, *Geschichtliche Grundbegriffe,* 1:xiii–xxvii.

——. "Historia Magistra Vitae: Über die Auflösung des Topos im Horizont neuzeitlich bewegter Geschichte." In *Natur und Geschichte: Karl Löwith zum 70. Geburtstag,* ed. Hermann Braun et al., 196–219. Stuttgart: Kohlhammer, 1967.

——. *Kritik und Krise: Eine Studie zur Pathogenese der bürgerlichen Welt.* 1959. Frankfurt: Suhrkamp, 1973.

——. *Zeitschichten: Studien zur Historik.* Frankfurt: Suhrkamp, 2003.

Koyré, Alexandre. *From the Closed World to the Infinite Universe.* Baltimore: Johns Hopkins P, 1957.

Landes, David S. *Revolution in Time: Clocks and the Making of the Modern World.* Cambridge, MA: Harvard UP, 1983.

Lang, Hans-Joachim. "Joel Barlows 'Dissertation on the Genius and Institutions of Manco Capac' als historisches Exemplum." In Engler and Müller, *Exempla,* 215–47.

Le Goff, Jacques. *Pour un autre Moyen Age: Temps, travail et culture en Occident; 18 essais.* Paris: Gallimard, 1977.

Lehmann, Karl. *Thomas Jefferson: American Humanist.* 1947. Chicago: U of Chicago P, 1965.

Lehmann, William L. *Henry Home, Lord Kames, and the Scottish Enlightenment: A Study in National Character and the History of Ideas.* The Hague: Martinus Nijhoff, 1971.

Lepenies, Wolf. *Das Ende der Naturgeschichte: Wandel kultureller Selbstverständlichkeiten in den Wissenschaften des 18. und 19. Jahrhunderts.* München: Hanser, 1976.

Levy, Leonard. *Jefferson and Civil Liberties: The Darker Side.* Cambridge, MA: Belknap of Harvard UP, 1963.

Lewis, Jan. " 'The Blessings of Domestic Society': Thomas Jefferson's Family and the Transformation of American Politics." In Onuf, *Jeffersonian Legacies,* 109–46.

Lewis, Jan, and Peter S. Onuf, eds. *Sally Hemings and Thomas Jefferson: History, Memory, and Civic Culture.* Charlottesville: UP of Virginia, 1999.

Lewis, Jan, et al. "Forum: Thomas Jefferson and Sally Hemings Redux." *William and Mary Quarterly* 57 (2000): 121–210.

Lienesch, Michael. *New Order of the Ages: Time, the Constitution, and the Making of Modern American Political Thought.* Princeton, NJ: Princeton UP, 1988.

Locke, John. *The Educational Writings of John Locke.* Ed. James L. Axtell. Cambridge: Cambridge UP, 1968.

——. *An Essay Concerning Human Understanding.* Ed. and abr. John W. Yolton. 1947. London: Dent, 2000.

Looney, J. Jefferson. " 'Beyond the Reach of Accident': Thomas Jefferson's Role in the Preservation of Virginia's Documentary Heritage." Unpublished research paper, 16 May 2008.

——. " 'I shall not retain a single one': The Limits of Thomas Jefferson's Library Catalogues." Paper presented at the conference "John Adams & Thomas Jefferson: Libraries, Leadership, and Legacy," Charlottesville, 26 June 2009.

Lovejoy, Arthur O. *The Great Chain of Being: A Study of the History of an Idea.* 1936. Cambridge, MA: Harvard UP, 1953.

Madison, James, Alexander Hamilton, and John Jay. *The Federalist Papers.* Ed. Isaac Kramnick. London: Penguin, 1987.

Malone, Dumas. *Jefferson and His Time.* 6 vols. Boston: Little, Brown, 1948–81.

Manning, Susan. *Fragments of Union: Making Connections in Scottish and American Writing.* Basingstoke: Palgrave, 2002.

———. *The Puritan Provincial Vision: Scottish and American Literature in the Nineteenth Century.* Cambridge: Cambridge UP, 1990.

Marshall, John. *The Life of George Washington.* 5 vols. 1804–7. Reprint, New York: AMS, 1969.

Martin, Edwin T. *Thomas Jefferson, Scientist.* New York: Henry Shuman, 1952.

McColley, Robert. *Slavery and Jeffersonian Virginia.* Urbana: U of Illinois P, 1964.

McCoy, Drew R. *The Elusive Republic: Political Economy in Jeffersonian America.* Chapel Hill: U of North Carolina P, 1980.

McDonald, Robert M. S. "Thomas Jefferson's Changing Reputation as Author of the 'Declaration of Independence': The First Fifty Years." *Journal of the Early Republic* 19 (1999): 169–95.

McLaughlin, Jack. *Jefferson and Monticello: The Biography of a Builder.* New York: Holt, 1988.

———. "Jefferson, Poe, and Ossian." In "Thomas Jefferson, 1743–1993: An Anniversary Collection." Special issue, *Eighteenth-Century Studies* 26, no. 4 (1993): 627–34.

Megill, Allan. "Aesthetic Theory and Historical Consciousness in the Eighteenth Century." *History and Theory* 17, no. 1 (1978): 29–62.

Miller, Charles A. *Jefferson and Nature: An Interpretation.* Baltimore: Johns Hopkins UP, 1988.

———. *Ship of State: The Nautical Metaphors of Thomas Jefferson, with Numerous Examples by Other Writers from Classical Antiquity to the Present.* Lanham, MD: UP of America, 2003.

Miller, John C. *The Wolf by the Ears: Thomas Jefferson and Slavery.* 1977. Charlottesville: UP of Virginia, 1991.

Momigliano, Arnaldo. *The Classical Foundations of Modern Historiography.* Berkeley and Los Angeles: U of California P, 1990.

Montesquieu, Charles-Louis de Secondat, Baron de la Brède et de. *Œuvres complètes.* Ed. Roger Caillois. 2 vols. Paris: Gallimard, 1951.

Nash, Gary B. *Race and Revolution.* Madison, WI: Madison House, 1990.

Newman, Richard S. *The Transformation of American Abolitionism: Fighting Slavery in the Early Republic.* Chapel Hill: U of North Carolina P, 2002.

Newton, Isaac. *Mathematical Principles of Natural Philosophy.* Chicago: Encyclopaedia Britannica, 1952.

Nicolaisen, Peter. "Thomas Jefferson, Sally Hemings, and the Question of Race: An Ongoing Debate." *Journal of American Studies* 37 (2003): 99–118.

O'Brien, Karen. *Narratives of Enlightenment: Cosmopolitan History from Voltaire to Gibbon.* Cambridge: Cambridge UP, 1997.

O'Malley, Michael. *Keeping Watch: A History of American Time.* New York: Viking, 1990.

Onuf, Peter S. "Domesticating the Captive Nation: Thomas Jefferson and the Problem of Slavery." In *Jefferson, Lincoln, and Wilson: The American Dilemma of Race and Democracy,* ed. John Milton Cooper Jr. and Thomas J. Knock. Charlottesville: U of Virginia P, 2010.

——. "Every Generation is an 'Independant Nation': Colonization, Miscegenation and the Fate of Jefferson's Children." *William and Mary Quarterly* 57 (2000): 153-70.

——, ed. *Jeffersonian Legacies.* Charlottesville: UP of Virginia, 1993.

——. *Jefferson's Empire: The Language of American Nationhood.* Charlottesville: UP of Virginia, 2000.

——. "The Scholars' Jefferson." *William and Mary Quarterly* 50 (1993): 671–99.

——. " 'To declare them a free and independant people': Race, Slavery, and National Identity in Jefferson's Thought." *Journal of the Early Republic* 18 (1998): 1–47.

Onuf, Peter S., and Nicholas G. Onuf. *Nations, Markets, and War: Modern History and the American Civil War.* Charlottesville: U of Virginia P, 2006.

Padgett, Alan G. *God, Eternity, and the Nature of Time.* New York: St. Martin's, 1992.

Paine, Thomas. *Thomas Paine: Collected Writings.* New York: Library of America, 1995.

Parton, James. *Life of Thomas Jefferson, Third President of the United States.* Boston: Osgood, 1874.

Pasley, Jeffrey L., Andrew W. Robertson, and David Waldstreicher, eds. *Beyond the Founders: New Approaches to the Political History of the Early American Republic.* Chapel Hill: U of North Carolina P, 2004.

Peden, William. Introduction to *Notes on the State of Virginia,* by Thomas Jefferson, ed. William Peden, xi–xxv. 1954. Chapel Hill: U of North Carolina P, 1982.

Peterson, Merrill D. *The Jefferson Image in the American Mind.* New York: Oxford UP, 1960.

——. *Thomas Jefferson and the New Nation: A Biography.* New York: Oxford UP, 1970.

Pocock, J. G. A. *Barbarism and Religion.* Vol. 2, *Narratives of Civil Government.* Cambridge: Cambridge UP, 1999.

——. *The Machiavellian Moment: Florentine Political Thought and the Atlantic Republican Tradition.* Princeton, NJ: Princeton UP, 1975.

Poe, Edgar Allan. *The Complete Poems and Stories with Selections from His Critical Writings.* Ed. Arthur Hobson Quinn. 2 vols. New York: Knopf, 1958.

Porter, Roy. *Enlightenment: Britain and the Creation of the Modern World.* London: Penguin, 2000.

Poulet, Georges. *Etudes sur le temps humain.* 2 vols. Paris: Plon, 1949.

Richard, Carl J. *The Founders and the Classics: Greece, Rome, and the American Enlightenment.* Cambridge, MA: Harvard UP, 1994.

Ricoeur, Paul. *Zeit und Erzählung.* Trans. Rainer Rochlitz. 3 vols. Munich: Fink, 1988-91.

Riesman, David. *The Lonely Crowd: A Study of the Changing American Character.* 1950. New Haven, CT: Yale UP, 1961.

Rifkin, Jeremy. *Time Wars: The Primary Conflict in Human History.* New York: Holt, 1987.

Ritter, Henning. "Du sollst deiner eigenen Zeit angehören." *Frankfurter Allgemeine Zeitung,* 25 September 2004, 37.

Rosenberger, Francis Coleman, ed. *Jefferson Reader: A Treasury of Writings about Thomas Jefferson.* New York: Dutton, 1953.

Ross, Dorothy. "Historical Consciousness in Nineteenth-Century America." *American Historical Review* 89, no. 4 (1984): 909–28.

Rossignol, Marie-Jeanne. *The Nationalist Ferment: The Origins of U.S. Foreign Policy, 1789–1812.* Trans Lillian A. Parrott. Columbus: Ohio State UP, 2004.

Rousseau, Jean-Jacques. *Discours sur l'origine et les fondements de l'inégalité parmi les hommes.* Paris: Editions Sociales, 1971.

Sawilla, Jan M. " 'Geschichte': Ein Produkt der deutschen Aufklärung? Eine Kritik an R. Kosellecks Begriff des Kollektivsingulars Geschichte." *Zeitschrift für historische Forschung* 31, no. 3 (2004): 381–428.

Schinkel, Anders. "Imagination as a Category of History: An Essay Concerning Koselleck's Concepts of Erfahrungsraum and Erwartungshorizont." *History and Theory* 44, no. 1 (2005): 42–54.

Schloss, Dietmar. *Die tugendhafte Republik: Politische Ideologie und Literatur in der amerikanischen Gründerzeit.* Heidelberg: Winter, 2003.

Schweizer, Niklaus R. *The Ut pictura poesis Controversy in Eighteenth-Century England and Germany.* Frankfurt: Peter Lang, 1972.

Seeber, Edward D. *Anti-Slavery Opinion in France during the Eighteenth Century.* 1937. New York: Burt Franklin, 1971.

Seifert, Arno. " 'Verzeitlichung': Zur Kritik einer neueren Frühneuzeitkategorie." *Zeitschrift für historische Forschung* 10, no. 4 (1983): 447–77.

Shackelford, George G. *Thomas Jefferson's Travels in Europe, 1784–1789.* Baltimore: Johns Hopkins UP, 1995.

Shaffer, Arthur H. *The Politics of History: Writing the History of the American Revolution.* Chicago: U of Chicago P, 1975.

Shalev, Eran. *Rome Reborn on Western Shores: Historical Imagination and the Creation of the American Republic.* Charlottesville: U of Virginia P, 2009.

Shapin, Stephen. *A Social History of Truth: Civility and Science in Seventeenth-Century England.* Chicago: U of Chicago P, 1994.

Sheehan, Bernard. *Seeds of Extinction: Jeffersonian Philanthropy and the American Indian.* Chapel Hill: U of North Carolina P, 1973.

Shelley, Percy Bysshe. "Time." In *The Poetical Works of Percy Bysshe Shelley,* ed. Thomas Hutchinson, 631. London: Frowde, 1905.

Sheridan, Eugene R. Introduction to *Jefferson's Extracts from the Gospels: "The Philosophy of Jesus" and "The Life and Morals of Jesus,"* ed. Dickinson W. Adams, 3–42. Princeton, NJ: Princeton UP, 1983.

Shuffleton, Frank, ed. *Thomas Jefferson: A Comprehensive, Annotated Bibliography of Writings about Him, 1826–1997.* http://etext.virginia.edu/jefferson/bibliog/.

Sloan, Herbert E. "The Earth Belongs in Usufruct to the Living." In Onuf, *Jeffersonian Legacies,* 281–315.

——. "Presidents as Historians." In *John Adams and the Founding of the Republic,* ed. Richard Alan Ryerson, 266–83. Boston: Massachusetts Historical Society, 2001.

——. *Principle and Interest: Thomas Jefferson and the Problem of Debt.* New York and Oxford: Oxford UP, 1995.

Smith, Adam. *The Theory of Moral Sentiments.* Ed. Knud Haakonssen. Cambridge: Cambridge UP, 1999.

——. *The Wealth of Nations.* Ed. Andrew Skinner. 2 vols. London: Penguin Classics, 1999.

Smith, Mark M. *Mastered by the Clock: Time, Slavery, and Freedom in the American South.* Chapel Hill: U of North Carolina P, 1997.

Sobel, Mechal. *The World They Made Together: Black and White Values in Eighteenth-Century Virginia.* Princeton, NJ: Princeton UP, 1987.

Sowerby, E. Millicent, comp. *Catalogue of the Library of Thomas Jefferson.* 5 vols. Washington, DC: Library of Congress, 1952–59.

Spahn, Hannah. " 'The Silent Course of Happiness': Domesticity and Politics in Jefferson's Presidency." In *The American Presidency: Multidisciplinary Perspectives,* ed. Dietmar Schloss, Martin Thunert, and Wilfried Mausbach. Heidelberg: Winter, 2011.

——. *Thomas Jefferson und die Sklaverei: Verrat an der Aufklärung?* Berliner Beiträge zur Amerikanistik. Ed. Winfried Fluck. Vol. 12. Berlin: John F. Kennedy-Institut für Nordamerikastudien, 2002.

Spahn, Peter. "Aristoteles." In *Pipers Handbuch der politischen Ideen,* ed. Iring Fetscher and Herfried Münkler, 397–437. Munich: Piper, 1988.

——. *Mittelschicht und Polisbildung.* Frankfurt: Peter Lang, 1977.

Stanton, Lucia. " 'Those Who Labor for My Happiness': Thomas Jefferson and His Slaves." In Onuf, *Jeffersonian Legacies,* 147–80.

Stein, Susan R. *The Worlds of Thomas Jefferson at Monticello.* New York: Abrams, 1993.

Sterne, Laurence. *The Florida Edition of the Works of Laurence Sterne.* Ed. Melvyn New et al. 8 vols. Gainesville: UP of Florida, 1978–2009.

——. *The Life and Opinions of Tristram Shandy, Gentleman.* Ed. Melvyn New et al. London: Penguin Classics, 1997.

——. *A Sentimental Journey through France and Italy. By Mr. Yorick.* Ed. Ian Jack. New York: Oxford UP, 1968.

Thompson, E. P. "Time, Work-Discipline, and Industrial Capitalism." In *Customs in Common,* 352–403. New York: New York UP, 1991.

Tocqueville, Alexis de. *Democracy in America.* Trans. Arthur Goldhammer. New York: Library of America, 2004.

Trevor-Roper, Hugh. "The Historical Philosophy of the Enlightenment." *Studies on Voltaire & the Eighteenth Century* 27 (1963): 1667–87.

Turetzky, Philip. *Time.* London: Routledge, 1998.

Tuveson, Ernest L. *Redeemer Nation: The Idea of America's Millennial Role.* Chicago: U of Chicago P, 1968.

Valsania, Maurizio. *The Limits of Optimism: Thomas Jefferson and the Drama of Enlightenment.* Charlottesville: U of Virginia P, 2011.

Vidler, Anthony. "Grégoire, Lenoir et les 'monuments parlants.' " In *La carmagnole des muses: L'homme de lettres et l'artiste dans la Révolution,* ed. Jean-Claude Bonnet, 131–54. Paris: Armand Colin, 1988.

Virgil [Publius Vergilius Maro]. *Aeneis.* German trans. August Vezin. Münster: Aschendorff, 1956.

Wallace, Anthony F. C. *Jefferson and the Indians: The Tragic Fate of the First Americans.* Cambridge, MA: Belknap of Harvard UP, 1999.

Walton, Craig. "Hume and Jefferson on the Uses of History." In *Hume: A Re-Evaluation,* ed. Donald W. Livingston and James T. King, 389–403. New York: Fordham UP, 1976.

Weber, Max. "Die Protestantische Ethik und der Geist des Kapitalismus." In *Gesammelte Aufsätze zur Religionssoziologie I.* 1920. Tübingen: Mohr, 1988.

Wellberg, David E. *Lessing's Laokoon: Semiotics and Aesthetics in the Age of Reason.* Cambridge: Cambridge UP, 1984.

Wendorff, Rudolf. *Zeit und Kultur: Geschichte des Zeitbewußtseins in Europa.* Opladen: Westdeutscher Verlag, 1980.

White, Hayden. *Metahistory: The Historical Imagination in Nineteenth-Century Europe.* Baltimore: Johns Hopkins UP, 1973.

Whitrow, G. J. *Time in History: Views of Time from Prehistory to the Present Day.* Oxford: Oxford UP, 1989.

Wiecek, William M. *The Sources of Antislavery Constitutionalism in America, 1760–1848.* Ithaca, NY: Cornell UP, 1977.

Wilcox, Donald J. *The Measure of Times Past: Pre-Newtonian Chronologies and the Rhetoric of Relative Time.* Chicago: U of Chicago P, 1987.

Williams, James G. "Proverbs and Ecclesiastes." In *The Literary Guide to the Bible,* ed. Robert Alter and Frank Kermode, 263–82. Cambridge, MA: Belknap of Harvard UP, 1987.

Wills, Garry. *Mr. Jefferson's University.* Washington, DC: National Geographic, 2002.

——. *"Negro President": Jefferson and the Slave Power.* Boston: Houghton Mifflin, 2003.

Wilson, Douglas L. "Jefferson vs. Hume." *William and Mary Quarterly* 46 (1989): 49–70.

Wilson, Gaye. "'Behold me at length on the vaunted scene of Europe': Thomas Jefferson and Creating an American Image Abroad." In *Old World, New World: America and Europe in the Age of Jefferson,* ed. Leonard Sadosky, Peter Nicolaisen, Peter S. Onuf, and Andrew J. O'Shaughnessy, 155–78. Charlottesville: U of Virginia P, 2010.

Wood, Gordon S. "Conspiracy and the Paranoid Style: Causality and Deceit in the Eighteenth Century." *William and Mary Quarterly* 39 (1982): 401–41.

——. *The Radicalism of the American Revolution.* New York: Knopf, 1992.

——. *Revolutionary Characters: What Made the Founders Different.* New York: Penguin, 2006.

——. "The Trials and Tribulations of Thomas Jefferson." In Onuf, *Jeffersonian Legacies,* 395–417.

Zerubavel, Eviatar. *The Seven Day Circle: The History and Meaning of the Week.* 1985. Chicago: U of Chicago P, 1989.

Ziesche, Philipp. *Cosmopolitan Patriots: Americans in Paris in the Age of Revolution.* Charlottesville: U of Virginia P, 2010.

Zilversmit, Arthur. *The First Emancipation: The Abolition of Slavery in the North.* Chicago: U of Chicago P, 1967.

Index

Italicized page numbers refer to illustrations.

Recent Books in the Jeffersonian America Series

Douglas Bradburn
The Citizenship Revolution: Politics and the Creation of the American Union, 1774–1804

Clarence E. Walker
Mongrel Nation: The America Begotten by Thomas Jefferson and Sally Hemings

Timothy Mason Roberts
Distant Revolutions: 1848 and the Challenge to American Exceptionalism

Peter J. Kastor and François Weil, editors
Empires of the Imagination: Transatlantic Histories of the Louisiana Purchase

Eran Shalev
Rome Reborn on Western Shores: Historical Imagination and the Creation of the American Republic

Leonard J. Sadosky
Revolutionary Negotiations: Indians, Empires, and Diplomats in the Founding of America

Philipp Ziesche
Cosmopolitan Patriots: Americans in Paris in the Age of Revolution

Leonard J. Sadosky, Peter Nicolaisen, Peter S. Onuf, and Andrew J. O'Shaughnessy, editors
Old World, New World: America and Europe in the Age of Jefferson

Sam W. Haynes
Unfinished Revolution: The American Republic in a British World, 1815–1850

Michal Jan Rozbicki
Culture and Liberty in the Age of the American Revolution

Ellen Holmes Pearson
Remaking Custom: Law and Identity in the Early American Republic

Seth Cotlar
Tom Paine's America: The Rise and Fall of Transatlantic Radicalism

John Craig Hammond and Matthew Mason, editors
Contesting Slavery: The Politics of Bondage and Freedom in the New American Nation

Ruma Chopra
Unnatural Rebellion: Loyalists in New York City during the Revolution

Maurizio Valsania
The Limits of Optimism: Thomas Jefferson's Dualistic Enlightenment

Peter S. Onuf and Nicholas P. Cole, editors
Thomas Jefferson, the Classical World, and Early America

Hannah Spahn
Thomas Jefferson, Time, and History